Essays on Self-Reference

Essays on Self-Reference

Niklas Luhmann

Columbia University Press
NEW YORK

Columbia University Press
New York   Oxford
Copyright © 1990 Columbia University Press
All rights reserved

Library of Congress Cataloging-in-Publication Data

Luhmann, Niklas.
Essays on self-reference / Niklas Luhmann.
p.   cm.
Includes biliographical references.
ISBN 0-231-06368-7
1. Social systems.   2. Autopoiesis.
3. Communication—Social aspects.   I. Title.
HM131.L817   1990   89-23905
302.2—dc20   CIP

Casebound editions of Columbia University Press books are Smyth-sewn and printed on permanent and durable acid-free paper

Printed in the United States of America

c 10 9 8 7 6 5 4 3 2 1

Contents

1. The Autopoiesis of Social Systems   1
2. Meaning as Sociology's Basic Concept   21
3. Complexity and Meaning   80
4. The Improbability of Communication   86
5. Modes of Communication and Society   99
6. The Individuality of the Individual: Historical Meanings and Contemporary Problems   107
7. Tautology and Paradox in the Self-Descriptions of Modern Society   123
8. Society, Meaning, and Religion—Based on Self-Reference   144
9. The State of the Political System   165
10. The World Society as a Social System   175
11. The Work of Art and the Self-Reproduction of Art   191
12. The Medium of Art   215
13. The Self-Reproduction of Law and Its Limits   227

Essays on Self-Reference

1.
The Autopoiesis of Social Systems

The term *autopoiesis* has been invented as a definition of life. Its origin is clearly biological. Its extension to other fields has been discussed, but rather unsuccessfully and on the basis of wrong premises. The problem may well be that we use a questionable approach to the problem, "tangling" our "hierarchies" of investigation.

At first sight it seems safe to say that psychic systems and even social systems are living systems too. Would there be consciousness or social life without life? And then, if life is defined as autopoiesis, how could one refuse to describe psychic systems and social systems as autopoietic systems? In this way we can retain the close relation of autopoiesis and life and apply this concept to psychic systems and to social systems as well. We are almost forced to do it, forced by our conceptual approach.[1] However, we immediately get into trouble in precisely defining what the "components" of psychic and social systems are whose reproduction by the same components of the same system recursively defines the autopoietic unity of the system. And what does "closure" mean in the case of psychic and social systems if our theoretical approach requires the inclusion of cells, neurophysiological systems, immune systems, etc. of living bodies into the encompassing (?) psychological or sociological realities?

Moreover, tied to life as a mode of self-reproduction of autopoietic systems, the theory of autopoiesis does not really attain the level of general systems theory which includes brains and machines, psychic systems and social systems, societies and short-term interactions. From this point of view, living systems are a special type of systems. If we abstract from life and define autopoiesis as a general form of system building using self-referential closure, we would have to admit that there are nonliving autopoietic systems, different modes of autopoietic reproduction, and that there are general principles of autopoietic organization that materialize as life, but also in other modes of circularity and self-reproduction. In other words, if we find nonliving autopoietic systems in our world, then and only then will we need a truly general theory of autopoiesis that carefully avoids references that hold true only for living systems. But which attributes of autopoiesis will remain valid on this highest level, and which have to be dropped on behalf of their connection with life?

The text that follows uses this kind of multilevel approach. It distinguishes a general theory of self-referential autopoietic systems and a more concrete level at which we may distinguish living systems (cells, brains, organisms, etc.), psychic systems, and social systems (societies, organizations, interactions) as different kinds of autopoietic systems. See figure 1.

Figure 1

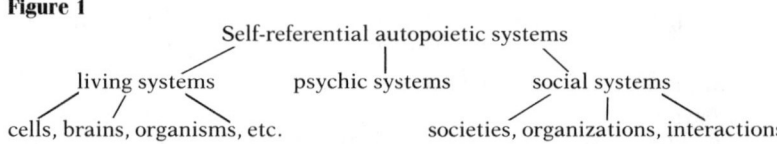

This scheme is not to be understood as describing an internal system differentiation. It is a scheme not for the operations of systems, but for their observation. It differentiates different types of systems or different modes of realization of autopoiesis.

This kind of approach is acceptable only if we are prepared to accept its anti-Aristotelian premise: that social systems and even psychic systems are not living systems. The concept of autopoietic closure itself requires this theoretical decision. It leads to a sharp distinction between meaning and life as different kinds of autopoietic organization, and meaning-using systems again have to be distinguished according to whether they use consciousness or communication as a mode of meaning-based reproduction. On the one hand, then, a psychological and sociological theory have to be developed that meet these requirements. On the other hand, the concept of autopoiesis has to be abstracted from biological connotations. Both tasks are clearly interdependent. The general theory of autopoietic systems founds the theories of psychic and social systems; the general theory itself, however, is meaningful only if this implementation succeeds, because otherwise we would be unable to determine which kind of attributes are truly general.

To use *ipsissima verba*: autopoietic systems "are systems that are defined as unities as networks of productions of components that recursively, through their interactions, generate and realize the network that produces them and constitute, in the space in which they exist, the boundaries of the network as components that participate in the realization of the network."[2] Autopoietic systems, then, are not only self-organizing systems, they not only produce and eventually change their own *structures;* their self-reference applies to the production of other *components* as well. This is the decisive conceptual innovation. It adds a turbocharger to the already powerful engine of self-referential machines. Even *elements*, that is, last components (in-dividuals) which are, at least for the system itself, undecomposable, are produced by the system itself. Thus, everything that is used as a unit by the system is produced as a unit by the system itself. This applies to elements, processes, boundaries, and other structures and, last but not least, to the unity of the system itself. Autopoietic systems, then, are sovereign with respect to the constitution of identities and differences. They, of course, do not create a material world of their own. They presuppose other levels of reality, as for example human life presupposes the small span of temperature in which water is liquid. But whatever they use as identities and as differences is of their own making. In other words, they cannot import identities and differences from the outer world; these are forms about which they have to decide themselves.

Social systems use communication as their particular mode of autopoietic reproduction. Their elements are communications that are recursively produced and reproduced by a network of communications and that cannot exist outside of such a network. Communications are not "living" units, they are not "conscious" units, they are not "actions." Their unity requires a synthesis of three selections, namely information, utterance,[3] and understanding (including misunderstanding).[4] This synthesis is produced by the network of communication, not by some kind of inherent power of consciousness, nor by the inherent quality of the information. Information, utterance, and understanding are aspects that for the system cannot exist independently of the system; they are co-created within the process of communication. Even "information" is not something that the system takes in from the environment.

Information doesn't exist "out there," waiting to be picked up by the system. As selection it is produced by the system itself in comparison with something else (e.g., in comparison with something that could have happened).

The communicative synthesis of information, utterance, and understanding is possible only as an elementary unit of an ongoing social system. As an operating unit it is undecomposable, doing its autopoietic work only as an element of the system. However, further units of the same system can distinguish between information and utterance and can use this distinction to separate hetero-referentiality and self-referentiality. They can, being themselves undecomposable for the moment, refer primarily to the content of previous communications, asking for further information about the information; or they can question the "how" and the "why" of the communication, focussing on its utterance. In the first case, they will pursue hetero-referentiality, in the second case self-referentiality. Using a terminology proposed by Gotthard Günther,[5] we can say that the process of communication is not simply auto-referential in the sense that it is what it is. It is forced by its own structure to separate and to recombine hetero-referentiality and self-referentiality. Referring to itself, the process has to distinguish information and utterance and to indicate which side of the distinction is supposed to serve as the base for further communication. Therefore, self-reference is nothing but reference to this distinction between hetero-reference and self-reference. And whereas auto-referentiality could be seen as a one-value thing and could be described by a logic with two values only, the case of social systems is a case of much higher complexity because its self-reference (1) is based on an ongoing auto-referential (autopoietic) process which refers to itself (2) as processing the distinction between itself (3) and its topics. If such a system didn't have an environment it would have to invent it as the horizon of its hetero-referentiality.

The elementary, undecomposable units of the system are communications of minimal size. This minimal size again cannot be determined independently from the system.[6] It is constituted by further communication or by the prospect of further communication. An elementary unit has the minimal meaning that is necessary for references by another communication, for instance the minimal meaning that still can be negated. Further communication can very well separate information, utterance, and understanding and discuss them as such but this still would presuppose their synthesis in previous communication. In a way, the system does not limit itself by using constraints for the constitution of its elementary units. If need be, it can communicate about everything and can decompose aspects of previous communication to satisfy actual desires. As an operating system, however, it will not always do this to an extreme.

Social systems, then, are recursively closed systems with respect to communications. However, there are two different meanings of "closure" that make it possible to distinguish *societies* and *interactions* as different types of social systems. Societies are encompassing systems in the sense that they include all events that, for them, have the quality of communication. They cannot communicate with their environment because this would mean including their understanding partner in the system, understanding being an essential aspect of the communication itself.[7] By communication they extend and limit the societal system, deciding about whether and what to communicate, and what to avoid. Interactions, on the other side, constitute their boundaries by the presence of people and are well aware that communication goes on around them without contact with their own actual interaction. They have to take into account environmental communication and they have to acknowledge the fact that the persons who are present and participate in the interaction have other roles and other obligations within systems that cannot be controlled here and now. But interactions also are closed systems in the sense that their own communication can be motivated and understood only in the context of the system, and if somebody approaches the interactional space and begins to participate, he has to be introduced and the topics of conversation eventually have to be adapted to the new situation. Interactions, moreover, cannot import communication ready made from their environment. They communicate or they don't communicate, according to whether they decide to reproduce or not to reproduce their own elements. They continue or discontinue their autopoiesis like living systems that continue as living systems or die. There are no third possibilities, neither for life nor for communication. All selections have to be adapted to the maintenance of autopoietic reproduction. Something has to be said, or at least good and peaceful (or bad and aggressive) intentions have to be shown if others are present.[8] Everything else remains a matter of structured choice within the system. Some of its structures, then, become specialized in assuring that communication goes on even if nothing of informative quality remains to be said[9] and even if the communication becomes controversial and nasty.

Confronted with the question of elementary units, most sociologists would come up with the answer: action. Sometimes "roles" or even human individuals are preferred. Since Max Weber and Talcott Parsons, however, action theory seems to offer the most advanced conceptualization.[10] Communication is introduced as a kind of action—e.g., as "kommunikatives Handeln" in the sense of Jürgen Habermas.[11] Usually this conceptualization is taken for granted and classical sociological theory finds itself resumed under the title of "Theory of Action."[12] Controversies are fought out under headings like "action versus system" or "individualistic versus holistic" approaches to social reality. There is no serious conceptual discussion that treats the relation of actions and communications, and the important question of whether action or communication should be considered as the basic and undecomposable unit of social systems has not been taken up.

For a theory of autopoietic systems, only communication is a serious candidate for the position of the elementary units of the basic self-referential process of social systems. Only communication is necessarily and inherently social; action is not. Moreover, social action implies communication, implies at least the communication of the meaning of the action or the intent of the actor; but it also implies the communication of the definition of the situation, of the expectation of being understood and accepted, and so on and so forth. And above all, communication is not a kind of action because it always contains a far richer meaning than uttering or sending messages alone. As we have seen, the perfection of communication implies understanding, and understanding is not part of the activity of the communicator and cannot be attributed to him. Therefore, the theory of autopoietic social systems requires a conceptual revolution within sociology: the replacement of action theory by communication theory as characterization of the elementary operative level of the system.

The relation of action and communication has to be reversed. Social systems are not composed of actions of a special kind, i.e., communicative actions, but require the attribution of actions to move on their own autopoiesis. Not psychological motivation and not reasoning or capacity of argumentation constitutes action but the attribution as such, that is, the linking of selection and responsibility for the narrowing of choice.[13] Only by attributing the responsibility for selecting the communication can the process of further communication be directed. One has to know who said what to be able to decide about further contributions to the process. Only by using this kind of simplifying localization of decision points can the process return to itself and communicate about communication.

Reflexive communication is not only an occasional event, it is a continuing possibility being co-reproduced by the autopoiesis itself. Every communication has to anticipate this kind of recursive elaboration, questioning, denial, or correction, and has to preadapt to these future possibilities. Only in working out this kind of presumptive fitness can it become part of the autopoietic process. This, however, requires the allocation and distribution of responsibilities and this function is fulfilled by accounting for action. The process therefore produces a second version of itself as a chain of actions. Contrary to the nature of communication itself, which includes the selectivity of information and the selectivity of understanding and thereby constitutes its elements by overlapping and partial interpenetration, this action chain consists of clear-cut elements that exclude each other. Contrary to the underlying reality of communication, the chain of communicative actions can be seen and treated as asymmetric.

In this sense the constitution and attribution of actions serve as a *simplifying self-observation* of the communicative system. The system processes information but it takes responsibility only for the action part of this process, not for the information. It is congruent with the world, it is universally competent, it includes all exclusions and at the same time is a system within the world, and it is able to distinguish and observe and control itself. It is a self-referential system and, thereby, a totalizing system. It cannot avoid operating within a "world" of its own. Societies constitute worlds. Observing themselves, that is communicating about themselves, they cannot avoid using distinctions that differentiate the observing system from something else. Their communication observes itself within its world and describes the limitations of its own competence. It never becomes self-transcending.[14] It can never use operations outside its own boundaries. The boundaries themselves, however, are components of the system and cannot be taken as given by a preconstituted world.

All of this sounds paradoxical, and rightly so. Social systems as seen by an observer are paradoxical systems.[15] They include self-referential operations, not only as a condition of the possibility of their autopoiesis but also due to their self-observation. The distinction of communication and action and, as a result, the distinction of world and system, are operative requirements. The general theory of autopoietic systems postulates a clear distinction between autopoiesis and observation. This condition is fulfilled in the case of social systems as well. Without using this distinction, the system could not accomplish the self-simplification necessary for self-observation. Autopoiesis and observation, communication and attribution of action are not the same and can never fuse. Nevertheless, self-observation in this specific sense of describing itself as a chain of clear-cut and responsible actions is a prerequisite of autopoiesis as such. Without this technique of using a simplified model of itself, the system could not communicate about communication and could not select its basic elements in view of their capacity to adapt themselves to the requirements of autopoiesis. This particular constellation may not be universally valid for all autopoietic systems. In view of the special case of social systems, however, the general theory has to formulate the distinction of autopoiesis and observation in a way that does not exclude cases in which self-observation is a necessary requirement of autopoiesis as such.

Observing such systems under the special constraints of logical analysis, we have to describe them as paradoxical systems or as "entangled hierarchies." It is not the task of an external observation to deparadoxize the system and describe it in a way that is suitable for multilevel logical analysis.[16] The system deparadoxizes itself. This requires "undecidable" decisions. In the case of social systems these are decisions about the attribution of action. If desired, these decisions themselves can be attributed as actions, which again could be attributed as actions and so on and so forth in infinite regress. Logically, actions are always unfounded actions and decisions are decisions exactly because they contain an unavoidable moment of arbitrariness and unpredictability. But this does not lead into lethal consequences. The system learns its own habits of acting and deciding,[17] accumulating experiences with itself and consolidating, on the basis of previous actions, expectations concerning future actions (structures). The autopoiesis does not stop in face of logical contradictions. It jumps, provided only that possibilities of further communication are close enough at hand.

The formal definition of autopoiesis gives no indication about the span of time during which components exist. Autopoiesis presupposes a recurring need for renewal. On the biological level, however, we are induced to think about the processes of replacement of molecules within cells or the replacement of cells within organisms, postponing for some time the finally unavoidable decay. The prolongation of life seems to be a way of paying the cost of evolutionary improbability. All complex order seems to be wrested from decay.

This holds true for social systems as well, but with a characteristic difference. Conscious systems and social systems have to produce their own decay. They produce their basic elements, i.e., thoughts and communications, not as short-term states but as events that vanish as soon as they appear. Events too occupy a minimal span of time, a specious present, but their duration is a matter of definition and has to be regulated by the autopoietic system itself: events cannot be accumulated. A conscious system does not consist of a collection of all of its past and present thoughts, nor does a social system pile up all of its communications. After a very short time the mass of elements would be intolerably large and its complexity would be so high that the system would be unable to select a pattern of coordination and would produce chaos. The solution is to renounce all stability at the operative level of elements and to use events only. Thereby, the continuing dissolution of the system becomes a necessary cause of its autopoietic reproduction. The system becomes dynamic in a very basic sense. It becomes inherently restless. The instability of its elements is a condition of its duration.

All structures of social systems have to be based on this fundamental fact of vanishing events, disappearing gestures or words that are dying away.[18] Memory, and then writing, have their function in preserving—not the events, but their structure-generating power.[19] The events themselves cannot be saved; their loss is the condition of their regeneration. Thus, time and irreversibility are built into the system not only at the structural level but also at the level of its elements. Its elements are operations and there is no reasonable way to distinguish "points" and "operations." Disintegration and reintegration, disordering and ordering require each other, and reproduction comes about only by a recurring integration of disintegration and reintegration.

This theoretical shift from self-referential structural integration to self-referential constitution of elements has important consequences for conceiving system maintenance. Maintenance is not simply a question of replication, of cultural transmission, of reproducing the *same patterns* under similar circumstances, e.g., using forks and knives while eating and only while eating, but the primary process is the production of *next elements* in the actual situation, *and these have to be different from the previous one* to be recognizable as events. This does not exclude the relevance of preservable patterns; it even requires them for a sufficient quick recognition of next possibilities. However, the system maintains itself not by storing patterns but by producing elements, not by transmitting *memes* (units of cultural transmission analogue to *genes*)[20] but by recursively using events for producing events. Its stability is based on instability. This built-in requirement of discontinuity and newness amounts to a *necessity to handle and process information*, whatever the environment or the state of the system offers as occasions. Information is an internal change of state, a self-produced aspect of communicative events and not something that exists in the environment of the system and has to be exploited for adaptive or similar purposes.[21]

If autopoiesis bases itself on events, a description of the system needs not only one, but two dichotomies: the dichotomy of system and environment and the dichotomy of event and situation.[22] Both dichotomies are world formulas: system plus environment is one way, event plus situation is another way to describe the world, presupposing a system reference. If the system (or its observer) uses the dichotomy of event/situation, it can see the difference of system and environment as a structure of the situation, the situation containing not only the system, but also its environment from the point of view of an event. Processing information by producing events-in-situations, the system can orient itself to the difference of internal and external relevances. As horizon (Husserl) of events, the situation refers to the system, to the environment, and to its difference—but all of this selectively, using the limited possibilities to produce the next event as a guideline.[23] Thus, the double dichotomy describes the way in which the system performs the "reentry" of the difference of system and environment into the system (I shall return to this point later). On the other hand, the difference of system and environment structures the limitation of choice that is needed to enable the system to proceed from event-in-situation to event-in-situation.

Systems based on events need a more complex pattern of time. For them, time cannot be given as irreversibility only. Events are happenings that make a difference between a "before" and a "thereafter." Events can be identified and observed, anticipated and remembered only as such a difference. Their identity is difference. Their presence is a copresence of the before and the thereafter. They have, therefore, to present time within time and have to reconstruct temporality in terms of a shifting presence which has its quality as presence only due to the double horizons of past and future that accompany the presence on its way into the future.[24] On this basis conscious time binding can develop.[25] The duality of horizons doubles as soon as we think of a future present or a past present, both of which have their own future and their own past. The temporal structure of time repeats itself within itself and only this reflexivity makes it possible to renounce a stable and enduring presence.[26] By a slow process of evolution the semantics of time has adapted to these conditions. For a long time it has used a religious reservation—*aeternitas, aevum,* or copresence of God with all times—to avoid the complete historization of time. Only the modern society recognizes itself—and consequently all previous societies—as constituting its own temporality.[27] The structural differentiation of society as an autonomous autopoietic system requires the coevolution of corresponding temporal structures, and modern historicism is the well-known result.

These short remarks by no means exhaust the range of suggestions that the theory of social systems can contribute to the abstraction and refinement of the general theory of autopoietic systems.[28] I return to the main idea with the question: what is new about it, given a long tradition of thinking about *creatio continua,* continuance, duration, maintenance, etc.?[29] Since the end of the sixteenth century, the idea of self-maintenance has been used to displace teleological reasoning and to reintroduce teleology by the argument that the maintenance of the system is the goal of the system or the function of its structures and operations. It is no surprise therefore that the question has been thrown into this discussion of what is added by the theory of autopoiesis to this well-known and rather futile traditional conceptualization.[30] An easy answer would be to mention the sharp distinction between self-reference on the level of structures (self-organization) and self-reference on the level of basic operations or elements. Moreover, we could point to the epistemological consequences of distinguishing autopoiesis and observation, observing systems being themselves autopoietic systems. We have only to look at the consequences of an "event-structure" approach for sociological theory to be aware of new problems and new attempts at solution, compared to the Malinowski/Radcliffe Brown/Parsons level of previous controversies. There is, however, a further aspect which should be made explicit.

The theory of autopoietic systems formulates a situation of binary choice. A system either continues its autopoiesis or doesn't. There are no in-between states, no third states. A woman may be pregnant or not, she cannot be a little pregnant. This is true, of course, for "system maintenance" as well. So, superficial observers will find only the same tautology. The theory of autopoietic systems, however, has been invented for a situation in which the theory of open systems has become generally accepted. Given this historical context, the concept of autopoietic closure has to be understood as the recursively closed organization of an open system. It does not return to the old notion of closed (versus open) systems.[31] The problem, then, becomes to see how autopoietic closure is possible in open systems. The new insight postulates closure as a condition of openness, and in this sense the theory formulates limiting conditions for the possibility of components of the system. Components in general and basic elements in particular can be reproduced only if they have the capacity to link closure and openness. For biological systems this does not require an "awareness" of, or knowledge about, the environment. For meaning-based conscious and social systems the autopoietic mode of meaning gives the possibility of "reentry,"[32] i.e., of presenting the difference of system and environment within the system. This reentered distinction structures the elementary operations of these systems. In social, i.e., communicative systems, the elementary operation of communication comes about by an "understanding" of the distinction of "information" and "utterance." Information can refer to the environment of the system. The utterance, attributed to an agent as action, is responsible for the autopoietic regeneration of the system itself. In this way information and utterance are forced to cooperate, are forced into unity. The emergent level of communication presupposes this synthesis. Without the basic distinction of information and utterance as different kinds of selection, the understanding would not be an aspect of communication, it would be a simple perception.

Thus, a sufficiently differentiated analysis of communication can show how the recurrent articulation of closure and openness comes about. It is a constitutive necessity of an emergent level of communication. Without a synthesis of three selections—information, utterance, and understanding—there would be no communication but simply perception. By this synthesis, the system is forced into looking for possibilities of mediating closure and openness. In other words: communication is an evolutionary potential for building up systems that are able to maintain closure under the condition of openness. These systems face the continuing necessity to select meanings that satisfy these constraints. The result is our well-known society.

In addition, the concept of autopoietic closure makes it possible to understand the function of enforced binary choices. The system can continue its autopoiesis or it can stop it. It can continue to live, to produce conscious states, to communicate with the only alternative to come to an end.[33] There are, with respect to autopoiesis, no third states. This is a powerful technical simplification. On the other hand, the system lacks any self-transcending power. It cannot enact operations of the outer world. A social system can only communicate. A living system can only live. Its autopoiesis, as seen by an observer, may have a causal impact on its environment. But autopoiesis is production in the strict sense of a process that needs further causes, not produced by itself, to attain its effect. *The binary structure of autopoiesis seems to compensate for this lack of totality.* It substitutes a kind of "internal totality." To be or not to be, to continue the autopoiesis or not to, serves as an internal representation of the totality of possibilities. Everything that can happen is reduced for the system to one of these two states. The world, whatever this is, may be indifferent to this question. The system emerges by inventing this choice, which does not exist without it. The negative value is a value not of the world but of the system. But it helps to simplify the totality of all conditions to one decisive question of how to produce the next system state, the next element, the next communication under the constraints of a given situation. Even unaware of the outer world, the living system "knows" that it is still alive and chooses its operations in using life for reproducing life. A communicative system too can continue to communicate on the basis of the ongoing communication. This does not require any reliable knowledge about outside conditions but the distinction of system and environment as seen from the point of view of the system. The unity of the autopoietic system is the recursive processing of this difference of continuing or not, which reproduces the difference as a condition of its own continuity. Every step has its own selectivity in choosing autopoiesis instead of stopping it. This is not a question of preference, nor a question of goal attainment. Rather, it has to be conceived as a "code" of existence, code

14 AUTOPOIESIS OF SOCIAL SYSTEMS

taken as an artificial duplication of possibilities with the consequence that every element can be presented as a selection.

This may become more clear if we consider the case of social systems. Autopoiesis here means: to continue to communicate. This becomes problematic in face of two different thresholds of discouragement. The first tends to stop the process because the communication *has not been understood*. The second tends to stop the process because *the communication has been rejected*. These thresholds reinforce each other because understanding increases the chances of rejection.[34] It is possible to refrain from communication in face of these difficulties and this is a rather common solution for interaction systems, particularly under modern conditions of highly arbitrary interactions. The society, however, the system of all communications, cannot simply capitulate in face of these problems. It cannot stop all communications at once and decide to avoid any renewal.[35] The autopoiesis of society has invented powerful mechanisms to guarantee its continuity in the face of a lack of understanding and even in the face of open rejection. It continues by changing the interactional context or by reflexive communication. The process of communication returns to itself and communicates its own difficulties. It uses a kind of (rather superficial) self-control to become aware of serious misunderstandings and it has the ability to communicate the rejection and restructure itself around this "no." In other words, the process is not obliged to follow the rules of logic. It can contradict itself. The system that uses this technique does not finish its autopoiesis and does not come to an end; it reorganizes itself as *conflict* to save its autopoiesis. In the case of serious problems of understanding and apparent misunderstandings, social systems very often tend to avoid the burden of argumentation and reasoned discourse to reach consensus—very much to the affliction of Habermas. Instead, they tend to push the matter into rejection and to embark on the easy vessel of conflict.

However this may be, the communication of contradiction, controversy, and conflict seems to function as a kind of *immune system* of the social system.[36] It saves autopoiesis by opening new modes of communication outside of normal constraints. The law records experiences and rules for behaving under these abnormal conditions and, by some kind of epigenesis, develops norms for everyday behavior which help to anticipate the conflict and to preadapt to its probable outcome.[37] In a highly developed society we even find a functionally differentiated legal system which reproduces its own autopoietic unity. It controls the immune system of the larger societal system by a highly specialized synthesis of normative (not-learning) closure and cognitive (learning) openness.[38] The law at the same time increases the possibilities of conflict, complexifies the immune system, and limits its consequences. It cannot, of course, exclude conflicts outside of the law that may save the autopoiesis of communication at even higher costs.[39]

A final point of importance remains: the epistemological consequences of autopoietic closure. This problem too has to be discussed with respect to the historical situation of scientific evolution in which the theory of autopoietic systems seems to offer advantages.

For many decades scientific research has no longer operated under the guidance of an undoubted orthodoxy—be it a theory of cognition or a theory of science in particular. The universally accepted expedient is "pragmatism": the only criteria of truth and of progressive knowledge are the results. This is clearly a self-referential, circular argument, based on a denial of circularity in theory, and on its acceptance in practice. The avoidance of circularity becomes an increasingly desperate stance—a paradox that seems to indicate that the condemned solution, the paradox itself, is on the verge of becoming accepted theory.

One way to cope with this ambiguous situation is to test methodologies with respect to their capacity to survive the coming revolution. Functional analysis is one of them. It can be applied to all problems, including the problem of paradox, circularity, undecidability, logical incompleteness, etc. Stating such conditions as a problem of functional analysis invites one to look for feasible solutions, for strategies of deparadoxization, of hierarchization (in the sense of the theory of types), of unfolding, of asymmetrization, etc, Functional analysis, in other words, reformulates the constitutive paradox as a "solved problem" (which is and is not a problem) and then proceeds to compare problem solutions.[40]

In addition to this kind of preadaptation in scientific evolution to an expected change of the paradigm of the theory of cognition itself, the theory of autopoietic systems constructs the decisive argument. It is a theory of a self-referential system applied to "observing systems" as well—"observing system" in the double sense that Heinz von Foerster chose as the title of a collection of his essays.[41] The theory distinguishes autopoiesis and observation, but it accepts the fact that observing systems themselves are autopoietic (at least: living) systems. Observation comes about only as an operation of autopoietic systems, be it life, consciousness, or communication. If it observes autopoietic systems it finds itself constrained by the conditions of autopoietic self-reproduction (again: life, respectively consciousness, respectively communication, e.g., language) and it includes itself in the fields of its objects, because as an autopoietic system observing autopoietic systems, it cannot avoid gaining information about itself.

In this way the theory of autopoietic systems integrates two separate developments of recent epistemological discussion. It uses a "natural" or even "material" epistemology, that is clearly distinct from all transcendental aspirations[42]—transcendentalism being in fact a title for the analysis of the autopoietic operations of conscious systems. In addition, it takes into account the special epistemological problems of universal or "global" theories, referring to a class of objects to which they themselves belong.[43] Universal theories, logic being one of them, have the important advantage of seeing *and comparing* themselves to other objects of the same type. In the case of logic, this would require a many-valued structure and the corresponding abstraction. The classical logic did not eliminate self-reference, but it had not enough space for its reflection. "The very fact that the traditional logic in its capacity of a place-value structure, contains *only* itself as a subsystem points to the specific and restricted role which reflection plays in the Aristotelian formalism. In order to become a useful theory of reflection a logic *has to encompass other sub-systems besides itself.*"[44] Only under this condition does functional analysis become useful as a technique of self-exploration of universal theories.

The usual objection can be formulated, following Nigel Howard, as the "existential axiom": knowing the theory of one's own behavior releases one from its constraints.[45] For an empirical theory of cognition, this is an empirical question. The freedom, gained by self-reflection, can be used only if its constraints are sufficiently close at hand. Otherwise the autopoietic system simply will not know what to do next. It may know, for example, that it operates under the spell of an "Oedipus complex" or a "Marxist" obsession, but it does not know what else it can do.

This kind of argument can generate biological,[46] psychological,[47] and sociological[48] epistemologies. Advances in substantial theory may have side effects on the theories that are supposed to control the research. Until the eighteenth century these problems were assigned to religion—the social system that specialized in tackling paradoxes.[49] We have retained this possibility, but the normalization of paradoxes in modern art and modern science seems to indicate our desire to eventually get along without religion.[50] Apparently our society offers the choice to trust religion or to work off our own paradoxes without becoming aware that this is religion.

Endnotes

1. See Humberto R. Maturana, "Man and Society," in Frank Benseler, Peter M. Hejl, and Wolfram Köck, eds., *Autopoiesis, Communication and Society: The Theory of Autopoietic System in the Social Sciences* (Frankfurt: 1980) pp. 11–31; and Peter M. Hejl, *Sozialwissenschaft als Theorie selbstreferentieller Systeme* (Frankfurt: 1982). See also Mario Bunge, "A Systems Concept of Society: Beyond Individualism and Holism," *Theory and Decision* (1979), 10:13–30.
2. Humberto R. Maturana, "Autopoiesis," in Milan Zeleny ed., *Autopoiesis: A Theory of Living Organization* (New York: North Holland, 1981), p. 21.
3. In German I could use the untranslatable term *Mitteilung*.
4. The source of this threefold distinction (which also has been used by Austin and Searle) is Karl Bühler, *Sprachtheorie: Die Darstellungsfunktion der Sprache* (Jena Fischer, 1934). However, I modify the reference of this distinction. It refers not to "functions" and not to types of "acts" but to selections.
5. See Gotthard Günther, Natural Numbers in Trans-Classic Systems, in Gotthard Günther, *Beiträge zur Grundlegung einer operationsfähigen Dialektik* (Hamburg: Meiner, 1979), 2:241–265.
6. This argument, of course, does not limit the analytical powers of an observer who, however, has to take into account the limitations of the system.
7. For problems of religion and particularly for problems of "communication with God" (revelation, prayer, etc.) see the essay "Society, Meaning, Religion—Based on Self-Reference."
8. This again is not a motive for action but a self-produced fact of the social system. If nobody is motivated to say something or to show his intentions, everybody would assume such communications and they would be produced without regard to such highly improbable psychological environment.
9. See Bronislaw Malinowski, "The Problem of Meaning in Primitive Language," in C. K. Ogden and J. A. Richards, eds., *The Meaning of Meaning: A Study of the Influence of Language upon Thought and of the Science of Symbolism*, 10th ed., 5th printing (London: 1960), pp. 296–336.
10. See the discussion of "The Unit of Action Systems" in Talcott Parsons, *The Structure of Social Action* (New York: 1937), pp. 43ff., which had an impact —one that is still apparent—on the whole theoretical framework of the later Parsons.
11. Jürgen Habermas, *Theorie des kommunikativen Handelns*, 2 vols. (Frankfurt: Suhrkamp, 1981).
12. See Richard Münch, *Theorie des Handelns: Zur Rekonstruktion der Beiträge von Talcott Parsons, Emile Durkheim und Max Weber* (Frankfurt: Suhrkamp, 1982).
13. To elaborate on this point I would have, of course, to distinguish "behavior" and "action." A corresponding concept of "motive" as a symbolic device facilitating the attribution of action has been used by Max Weber. See also C. Wright Mills, "Situated Actions and Vocabularies of Motive," *American Socio-*

logical Review (1940), 5:904–913; Kenneth Burke, *A Grammar of Motives* (1945) and *A Rhetoric of Motives* (1950) both reprinted; (Cleveland, Ohio: World Publishing Company, 1962); and Alan F. Blum and Peter McHugh, "The Social Ascription of Motives," *American Sociological Review* (1971), 36:98–109.

14. See the distinction between perceiving oneself and transcending oneself by Douglas R. Hofstadter, *Gödel, Escher, Bach: An Eternal Golden Braid* (Hassocks, Sussex, England: Harvester Press, 1979), p. 478.

15. The term *paradox* refers to a logical collapse of a multilevel hierarchy, not to a simple contradiction. See Anthony Wilden, *System and Structure: Essays in Communication and Exchange* (London: Tavistock, 1972), pp. 390ff; Hofstadter, *Gödel, Escher, Bach op. cit.*; see also Yves Barel, *Le paradoxe et le système: Essai sur le fantastique social* (Grenoble: Presses Universitaires, 1979).

16. I do not comment on the possibility of a logical analysis of self-referential systems that bypasses the Gödel limitations and avoids hierarchization.

17. "Learning" should be understood as an aspect of autopoiesis, i.e., as a change of structure within a closed system (and not: as adaption to a changing environment). See Humberto R. Maturana, "Reflexionen: Lernen oder ontogenetischer Drift," *Delfin* (1983) II:60–71.

18. It is rare that social scientists have a sense for the radicality and the importance of this insight. But see Floyd H. Allport, "An Event-System Theory of Collective Action: With Illustrations from Economic and Political Phenomena and the Production of War," *The Journal of Social Psychology* (1940), 11:417–445; "The Structuring of Events: Outline of a General Theory with Applications to Psychology," *The Psychological Review* (1954), 61:281–303.

19. This explains that the invention of writing speeds up the evolution of complex societal systems, making it possible to preserve highly diversified structural information. This is, by now, a well-explored phenomenon that still lacks a sufficient foundation in general theory. See Frances A. Yates, *The Art of Memory* (London: Routledge, 1966); Walter J. Ong, *The Presence of the Word: Some Prolegomena for Cultural and Religious History* (New Haven, Conn.: Yale University Press, 1967); Eric A. Havelock, *The Literate Revolution in Greece and Its Cultural Consequences*, (Princeton, N.J.: Princeton University Press, 1982).

20. Dawkins' term. See R. Dawkins, *The Selfish Gene* (New York: Oxford University Press, 1976).

21. See also (for systems) Heinz von Foerster, *Observing Systems* (Seaside, Calif.: Intersystems Publications, 1981), p. 263: "The environment contains no information; the environment is as it is."

22. Using this theoretical framework, one cannot speak of an "environment of events, of actions, etc." nor of "situations of a system."

23. See, for further elaborations using the method of phenomenological psychology, Jürgen Markowitz, *Die soziale Situation* (Frankfurt: Suhrkamp, 1979).

24. One of the best analyses of this complicated temporal structure remains Edmund Husserl, "Vorlesungen zur Phänomenologie des inneren Zeitbewuβtseins," *Jahrbuch für Philosophie und phänomenologische Forschung* (1928), 9:367–496. For social systems, see Werner Bergmann, *Die Zeitstruckturen sozialer Systeme: eine systemtheoretische Analyse* (Berlin: Duncker and Humblodt, 1981).

25. See Alfred Korzybski, *Science and Sanity: An Introduction to Non-aristotelic Systems and General Semantics*, 3d. ed. (reprint Lakeville, Conn: 1949). From an evolutionary point of view, see G. Ledyard Stebbins, *Darwin to DNA, Molecules to Humanity* (San Francisco, Calif.: Freeman, 1982), pp. 363f.

26. See Niklas Luhmann, "The Future Cannot Begin: Temporal Structures in Modern Society," in Niklas Luhmann, *The Differentiation of Society* (New York: Columbia University Press,1982), pp. 271–288.

27. See Niklas Luhmann, "World-Time and System History: Interrelations Between Temporal Horizons and Social Structures," in Luhmann, *The Differentiation of Society*, op. cit. pp. 289–323; Niklas Luhmann, "Temporalisierung von Komplexität: Zur Semantik neuzeitlicher Zeitbegriffe," in Luhmann, *Gesellschaftsstruktur und Semantik* (Frankfurt: Suhrkamp, 1980) 1:235–300.

28. For a more extensive treatment see Niklas Luhmann, *Soziale Systeme: Grundriß einer allgemeinen Theorie* (Frankfurt: Suhrkamp, 1984).

29. See Hans Ebeling, ed., *Subjektivität und Selbsterhaltung: Beiträge zur Diagnose der Moderne* (Frankfurt: Suhrkamp, 1976) and of course the extensive "functionalistic" discussion about "system maintenance."

30. See Erich Jantsch, "Autopoiesis: A Central Aspect of Dissipative Self-Organization," in Zeleny, *Autopoiesis*, pp. 65–88, preferring the theory of thermodynamic disequilibrium and dissipative structures.

31. See Francisco J. Varela, *Principles of Biological Autonomy* (New York; North-Holland, 1979).

32. Reentry in the sense of George Spencer Brown, *Laws of Form*, 2d ed. (London. Allen and Unwin, 1971). Gotthard Günther makes the same point in stating "that these systems of self-reflection with centers of their own could not behave as they do unless they are capable of 'drawing a line' between themselves and their environment." And this leads Günther "to the surprising conclusion that *parts of the universe have a higher reflective power than the whole of it.*" See "Cybernetic Ontology and Transjunctional Operations," in Gotthard Günther, *Beiträge zur Grundlegung einer operationsfähigen Dialektik* (Hamburg: Meiner, 1976), 1:319.

33. "End" in this theory, therefore, is not "telos" in the sense of the perfect state, but just the contrary: the zero state which has to be avoided by reproducing imperfect and improbable states. In a very fundamental way the theory has an anti-Aristotelian drift.

34. From an evolutionary point of view, see the essay, "The Improbability of Communication."

35. That the physical destruction of the possibility of communication has become possible and that this destruction can be intended and produced by communication is another question. In the same sense, life cannot choose to put an end to itself, but conscious systems can decide to kill their own bodies.

36. See Niklas Luhmann, *Soziale Systeme*, pp. 488ff.

37. See Niklas Luhmann, *A Sociological Theory of Law* (London: Routledge, 1985).

38. For further elaboration see Niklas Luhmann, "The Unity of the Legal System," in Gunther Teubner, ed., *Autopoietic Law* (Berlin: De Gruyter, 1988), pp. 12–35.

39. Recent tendencies to recommend and to domesticate symbolic illegalities as a kind of communication adapted to reacting against much integration of society and positive law seem to postulate a kind of second immune system on the base of a revived natural law, of careful choice of topics and highly conscientious practice. See Bernd Guggenberger, "An den Grenzen der Verfassung," *Frankfurter Allgemeine Zeitung* (December 3, 1983), no. 281.

40. In a way, the problem remains a problem by "oversolving" it, i.e., by

inventing several solutions that are of unequal value and differ in their appropriateness according to varying circumstances. This gives us, by a functional analysis of functional analysis, an example of how deparadoxization can proceed. The happy pragmatist, on the other hand, would be content with stating that a problem becomes a problem only by seeing a solution. See Larry Laudan, *Progress and Its Problems: Toward a Theory of Scientific Growth* (Berkeley: University of California Press, 1977).

41. Heinz von Foerster, *Observing Systems*.

42. See Willard van O. Quine, "Epistemology Naturalized," in: Willard van O. Quine, *Ontological Relativity and Other Essays* (New York: 1969), pp. 69–90.

43. See C. A. Hooker, "On Global Theories," *Philosophy of Science* (1975), 42:152–179. Many examples: the theory of sublimation may be itself a sublimation. Physical research uses physical processes. The theory of the self has to take into account that the theorist himself is a self (a healthy self, a divided self). For this last example see Ray Holland, *Self and Social Context* (New York: St. Martin's, 1977).

44. Gotthard Günther, "Cybernectic Ontology and Transjunctional Operations," in Gotthard Günther, *Beiträge zur Grundlegung einer operationsfähigen Dialektik* (Hamburg: Meiner, 1976), 1:310, my emphasis.

45. See Nigel Howard, *Paradoxes of Rationality: Theory of Metagames and Political Behaviour* (Cambridge, Mass.: MIT Press, 1971) pp. 2ff. and passim.

46. See Humberto R. Maturana, "Cognition," in Peter M. Hejl et. al., eds., *Wahrnehmung und Kommunikation* (Frankfurt: Lang, 1978), pp. 29–49.

47. See Donald T. Campbell, "Natural Selection as an Epistemological Model," in Raoul Naroll and Ronald Cohen, eds., *A Handbook of Method in Cultural Anthropology* (Garden City, N.Y.: The Natural History Press, 1970) pp. 51–85; Donald T. Campbell, "Evolutionary Epistemology," in Paul A. Schilpp, ed., *The Philosophy of Karl Popper* (La Salle, Ill.: 1974), pp. 413–465; Donald T. Campbell, "On the Conflicts Between Biological and Social Evolution and Between Psychological and Moral Tradition," *American Psychologist* (1975), 30:1103–1126.

48. For a case study, using this mode of controlled self-reference see Jonathan R. Cole and Harriet Zuckerman, "The Emergence of a Scientific Specialty: The Self-Exemplifying Case of the Sociology of Science." in Lewis A. Coser, ed., *The Idea of Social Structure: Papers in Honor of Robert K. Merton* (New York: 1975), pp. 139–174.

49. See the impasse as formulated by Bishop Huet: "Mais lors que l'Entendement en vue de cette Idée forme un jugement de l'objet exterieur, d'où cette Idée est partie, il ne peut pas savoir très certainement et très clairement si ce jugement convient avec l'objet exterieur; et c'est dans cette convenance que consiste la Verité, comme je l'ai dit. De sorte qu'encore qu'il connoisse la Verité, il ne scait pas qu'il connoît, et il ne peut être assuré de l'avoir connue." Pierre Daniel Huet, *Traité Philosophique de la Foiblesse de l'esprit humain* (Amsterdam: Du Sauzet, 1723), p. 180.

50. See Jean-Pierre Dupuy, *Ordres et Désordres: Enquête sur un nouveau paradigme* (Paris: Seuil, 1982) pp. 162ff.

2.
Meaning as Sociology's Basic Concept

The amount of attention that sociology devotes to its basic concepts remains rather modest. There may be reasons of method for this, or other grounds. Some theories of science, for example, posit a high, almost unrestricted freedom in the choice fundamental categories, with their value to be determined only by prognostic or explanatory success. This position properly discourages the naive discussion of concepts based on the assumption that there exist correct, or at least successful, concepts, which need only be unearthed and cleaned up a little. It has its own naiveté in the assumption that the real world has already been decided upon and that it remains only to determine the facts with the help of suitable concepts. A radical critique of conceptual realism, however, must also do away with cosmological realism. This was shown with unmistakable clarity by the course of late medieval—early modern philosophy. This is something we can know. Under these circumstances, sociology still has the option of basing its claim to being scientific on a denial of such knowledge and of seeing its positivity in just this assumption of a pregiven world. But it cannot prevent this position from representing a decision—a decision taken in the face of better knowledge. The obvious and unavoidable selectivity in the position of sociological positivism prompts us to reopen the discussion of basic categories in a new form and to proceed under the opposite premise, viz. that neither concepts nor the world can be treated simply as given. Such a premise would appear to lead to absurdities. But then absurd premises, in excluding nothing, do have the advantage of minimizing the chance of error.

What can no longer be presupposed will have to be brought forth

22 MEANING AS BASIC CONCEPT

in the construction of our basic categories. The basic concepts of a discipline that is willing to face up to the assumption of a contingent world must be able to deal with this problem. They must confront it. Their suitability will have to be judged using different criteria, i.e., no longer from the point of view of the accurate reproduction of what is simply pregiven and waiting to be discovered, but from that of grasping and reducing this contingency of possible worlds. As the basic category for describing how this is accomplished in consciousness and communication (and not merely physically or organically) I suggest the concept of *meaning*.

The concept of meaning has been well prepared for this function by the investigations of subjective-transcendental metaphysics, in particular its epistemology. Its explicit articulation within the framework of this tradition, however, involves theoretical assumptions—in particular that of objectively determinable rules or forms or values (or more recently: interests) underlying and guiding cognition or the acquisition of knowledge—that sociology can hardly accept if their transcendental status is to be retained. Neither does the content of this concept—as interpreted within this tradition—meet the demands that must be placed on one of sociology's fundamental categories.

The transcendental tradition suggested seeking a clarification of the concept of meaning by reference to a subject and defining meaning through subjective intention.[1] Meaning was seen as characterized by the conscious actualization of the intentional structures of experience, and accompanying this and available in reflection was a consciousness of the pregivenness and uniqueness (I-ness) of the experiencing subject. While such reflection does not have the intention of placing the subject outside of Being, it certainly does have this effect. It gives us something that cannot be: an isolated ego. The paradoxical nature of this result is blurred by the introduction of a number of distinctions, most notably that between the transcendental and empirical subject and that between meaning and Being. The function of these distinctions lay in the analysis (dissection) of a theoretical impasse, but the attempt can hardly be regarded as successful. The problem is found again in the concept pairs used to make these distinctions and reappears in the question of the reality of the transcendental (nonempirical) subject, in the impossibility of treating the empirically objectifiable "subject" as *subiectum (hypokeimenon)* of the world—i.e., of taking the subjectivity of this subject seriously—and, finally, in the impossibility of conceiving of meaning without being or being without meaning.

Once the chain of dependencies in concept formation is recognized, it becomes clear that the analysis must start before, and not after, this impasse—that it should not try to disentangle it but to avoid it in the first place. The source of the problem surely lies in an ill-defined concept of reflection, one that is far too vague and far too compact in its depiction of the concrete life of consciousness. A closer analysis of this life of consciousness would have to, at the very least, distinguish between, on the one hand, intentions that consciously constitute the self as a system-in-the-world and, on the other, the reflexivity of particular kinds of processes within consciousness, for example, the thinking of thought, the feeling of feeling, the willing of willing—which, while possible only with a higher specification of the processes in question, do make possible a more complex constitution of the self system.[2] In no case, however, should this impasse and all its attendant conceptual strategies continue to burden the concept of meaning and, through it, other sociological research concepts. This warning can be formulated more pointedly: the position I am criticizing here merely transfers the problem to the question of the meaning of the subject and leaves it there, unresolved. As undeniable as the relationship between meaning and consciousness is, its clarification must be undertaken the other way around. The concept of meaning is primary, and should be defined without reference to the concept of subject, since the latter—as a meaningfully constituted identity—already presupposes the concept of meaning.

I am going to replace reference to a subject with a much more highly differentiated analytical instrumentarium, in which the concepts of function and of system play a special role. I shall start with an analysis of the function of meaning and attempt to show that fulfilling this function presupposes the existence of meaning-constituting systems. I shall not try here to anticipate the results of this undertaking, but should nonetheless attempt to guard against certain misunderstandings that could crop up along the way. This will be best accomplished if we start with a short look at what the concept of meaning-constituting systems does, and does not, accomplish.

While reserving for later a closer characterization of the concepts of meaning and system, I can state here that when I refer to meaning-constituting systems I am not speaking of some source of energy or some kind of cause or of the psychic-organic bases of meaningful experience, and certainly not of the concrete individual, but rather of an interconnected complex of meaning *(Sinnzusammenhang)* as such. This includes both psychic systems—as far as they are identified (by anyone!) as the unity of a meaningfully related complex of actions and experiences—and social systems. In outlining my basic concepts here I am thus aiming at elements that are fundamental to both psychic and social systems alike and first make possible their differentiation. Concepts like experience and action, expectation and disappointment will also be defined in such a way that they do not necessarily carry a psychological meaning; the question of referring them to a psychic or to a social system (and thus of fitting them into the conceptual framework of psychology or sociology) is left open. Only when such a decision is made, i.e., only with the choice of system referents, does that which is designated by the terms *meaning, experience, action,* etc. become a psychological or a sociological category. This is, of course, not to deny that all meaning, all experience, all action presupposes psychic systems together with their organic substructure and is possible only on the basis of these. But I want to express this undeniable fact by saying that it is possible to specify a psychic system referent for any and all experience; for just the same can be said of social systems: without them, neither meaning nor action would be possible, so that here too we have no instance without some system. The advantage of this formulation is (a) that it counters the widespread notion that the foundation of meaning and so forth in psychic systems is somehow more fundamental, more originary, simpler, or more elemental than a foundation in social systems and (b) that it leaves open, i.e., leaves as a matter for investigation, the question of which choices of system referents are meaningful in which scientific or life-world contexts.

As long as we are dealing with topics for which system referents —although they must be chosen in the end—have not yet been specified, the methods of empirical science do not apply. I shall, instead, have to rely in this stage of the investigation on techniques like those specially developed in phenomenology (and, to a lesser extent, hermeneutics) for the analysis of meaning formations: techniques for the abstracting grasp of a given item of meaning, for its replication and variation in thought, for its construction as an alternative to other possibilities.

Apart from such abstractness and from my recourse to phenomenology (which some will regard more as a problem than a solution), what is alienating about the concept of meaning-constituting systems is just the most crucial aspect: the lack of clarity concerning the relationship between meaning and system. I am going to refer to this relationship as *constitution*—without wishing to claim that this word alone affords adequate clarification. What I am trying to get at here—and this will have to be looked at more closely below—is the fact that meaning always appears within some delimitable context and yet at the same time always points beyond this context and lets us see other possibilities. What I want to understand and to describe with the term *constitution* is this relationship between a selectively restricted order and the openness of other possibilities, a relationship of mutual dependence, of being-possible-only-together.[3] Taking up recent developments in systems theory, I am going to try—and this is the core of what follows—to interpret this typical constitution relationship for meaningful experience and action using the concepts of system and world (or environment) and will therefore be speaking of meaning systems as meaning-constituting systems.

We will have to bear in mind that the concept of meaning refers to the way human experience is ordered and not, for example, to some particular fact or matter in the world. A direct and presuppositionless approach to the problem of meaning is then best sought in a phenomenological description of what actually presents itself in meaningful experience.

If this question is approached with the openness for which Husserl's work is still a model, we will arrive at a final, incontestable elementary result: the momentary Given that fills experience at any time always and irrevocably refers beyond itself to something else. Experience experiences itself as variable—and unlike transcendental phenomenology we assume organic bases for this. It does not find itself closed and self-contained, not restricted to itself, but is always referring to something that is at that moment not its actual content. This referring-beyond-itself, this immanent transcendence of experience, is not a matter of choice; rather, it is the condition on the basis of which all freedom to choose must first be constituted. Nor is a reflexive turning to experience as such (let alone, to the experiencing subject) able to escape this condition; it too exhibits the same structure and simply points experience in a particular direction, alongside which others still remain possible. The problem of integrating the actuality of experience with the transcendence of its other possibilities remains inescapable, and inescapable, too, is the form of experience processing that accomplishes this. It is this that I call meaning. There is, then, if we follow this usage, no such thing as meaningless experience. The efforts some make today to confront us with meaninglessness ultimately serve only the increasingly difficult production of astonishment; and the positivist who labels a Christmas carol "meaningless" is merely formulating the limits of his own maxim. In its very strangeness all non-sense (this too) still has meaning.

In what exactly does this problem of integrating experience with its other, transcendent possibilities consist? A functional definition of the concept of meaning will afford an answer to this question.

The most important feature of the differentiation between Actuality and Potentiality found in experience resides in the character of the overabundance of possibilities, which by far exceeds what can be realized through action or actualized in experience. The actual, given contents of experience always point by way of reference and implication to far more—whether taken together or as a sequence—than can be brought into the narrow spotlight of consciousness. Alongside direct, immediate conscious experience there stands a world of other possibilities. This self-overburdening of experience with other possibilities exhibits the double structure of complexity and contingency. The term *complexity* is meant to indicate that there are always more possibilities of experience and action than can be actualized. The term *contingency* is intended to express the fact that the possibilities of further experience and action indicated in the horizon of actual experience are just that—possibilities—and might turn out differently than expected,[4] i.e., that these indications can be deceptive: perhaps they point to something that is not really there or cannot be reached in the way expected; perhaps even after the necessary steps have been taken (e.g., someone has gone to a particular place) what was expected can no longer be actualized, because events in the meantime have removed or destroyed this possibility. In practice, then, complexity means the necessity of choosing; contingency, the necessity of accepting risks.

Under this condition of complex and contingent other possibilities, experience takes the form of risk-laden selectivity. One may assume that this is true of organic life in general and conceive of the specifically human solution to this problem as a partly continuous, partly discontinuous improvement in this achievement.[5] It is a mark of conscious experience, as opposed to purely organic selection, that it regulates itself through just this self-overburdening, with complexity and contingency regulating the selective processing of experience in a very specific way, viz. in the form of meaning.

This requires that the complexity of other possibilities be constituted within experience itself and *remain preserved there*. Experience and action are unceasing selection but cannot simply eliminate those alternatives not chosen or make them totally disappear until chance brings them into view once more; instead, they can only neutralize them. Complexity cannot simply be "erased," as computer jargon puts it (this is adequate only for machines), but is, so to speak, only placed in brackets, reduced from moment to moment in continually different ways, and always remains preserved as the most generally constituted selection domain, as the "source" of constantly new and constantly different additional choices—as the world.

With these few remarks I have touched upon the central problem here, the problem with reference to which the concept of meaning may be given a functional definition. Meaning functions as the premise for experience processing in a way that makes possible a choice from among different possible states or contents of consciousness, and in this it does not totally eliminate what has not been chosen, but preserves it in the form of the world and so keeps it accessible.[6] The function of meaning then does not lie in information, i.e., not in the elimination of a system-relative state of uncertainty about the world, and it cannot, therefore, be measured with the techniques of information theory. If it is repeated, a message or piece of news loses its information value, but not its meaning. Meaning is not a selective event, but a selective relationship between system and world—although this is still not an adequate characterization. Rather, what is special about the meaningful or meaning-based processing of experience is that it makes possible *both* the reduction and the preservation of complexity; i.e., it provides a form of selection that prevents the world from shrinking down to just one particular content of consciousness with each act of determining experience.

The constitution of this world of not-yet-actualized potentialities, which is constantly given and accompanies all experience, rests on the specifically human capacity for *negation*. Its conceptual reconstruction will require clarity concerning *the functional primacy of negativity in meaning-constituting experience*. I shall leave open here the question of what negativity "in itself" may be.[7] Still, we must consider the question of whether, in the case of negativity, we are dealing with a qualitatively unified something, with an irreducible logical atom of meaning, or whether such an assumption is simply foisted upon us by language. A functional analysis of negating will be able to advance beyond merely tautological explanations only if we look more closely at *how* negating contributes to the constitution of meaning. And here we discover a complex network of achievements.[8]

The function of negation as a highly involved strategy of experience processing can be clarified enough for our purposes here. Negation appears not only to be the most universally usable language symbol but, in addition, to actually constitute the universality, i.e., the world reference, of all practical life—especially where experience or action is directed positively toward some particular meaning and intends it under the form of "is" or "ought." The specific potency of negation, something not to be found in the pure givenness of actual impressions, in perception, or imagination, stems from its own special combination of *reflexivity* and *generalization*. Negation is a reflexive (and a necessarily reflexive) process form of experience. It can be applied to itself, and this possibility of the negation of negation is indispensable in any experience that can negate at all. This, however, means that all negation remains irredeemably provisional and does not permanently block our access to what has been negated. Only time, not negation, eliminates possibilities definitively.

The reflexivity of negation both depends on and supports generalization. First and most important, whenever something is positively attended to, negation forms part of this experience, and affords it security. When I turn to one particular thing, I do so in the certainty that "everything else" remains preserved—both what is present, but momentarily not of interest, and what is absent, in particular the danger that is not-present and whose ongoing negatability first allows me to attend to something else. I specify my yes and leave the negations necessary for this unspecified. The high risks involved in such a wholesale bracketing out of what is not directly attended to are decidedly lessened by the element of preservation involved in all negation. I always reserve the right to negate such negations as the need arises and to attend positively to whatever unexpected problems present themselves. Only this reservation makes the wholesale exclusion of other possibilities practicable, for it allows it to be corrected on a case-by-case basis. Generalization and reflexivity thus function in a necessary relationship as mutually dependent components of negation.

On the other hand, the explicit negation of some particular meaning is a special case, likely made possible only by language. Direct and specific negation (a certain thing is not there; I am not going to do something) ultimately intends this preserving effect itself: it remains unspecific and indifferent (in this sense: generalized) with respect to what has taken, or could take, the place of the negated elements, and it serves—through learning[9] or the formation of norms—to maintain the overall structure of expectations out of which the one element has become problematic.[10]

I can summarize all of this and specify a little more closely a first and very basic aspect of the way meaningful or meaning-based experience processing functions: it achieves both the *reduction and preservation of complexity* by filling immediately given, evident experience with *references to other possibilities* and with a *reflexive and generalizing negation potential*, thus equipping it for *risk-laden selectivity*.

This way of looking at things does not eliminate all reference of the concept of meaning to "consciousness"—but it does alter its form. Consciousness is no longer regarded as the subject *(hypokeimenon, subiectum)* of meaning in the sense of something that can be discovered and substantialized in reflection, but rather as that experiencing with reference to which meaning can be functionally analyzed (and whose limits and potential we will have to examine).[11] If we wish to proceed from here to a theory of consciousness, it will not be enough to simply replace the old notion of consciousness as a picture of the real world with the concept of reflection—i.e., to conceive of consciousness (with Hegel and Fichte) as an act or a reflexive process. This does introduce the temporal dimension—a gain that cannot be relinquished—but the concepts of action and reflection used to accompany this imply far too simple a process, one that is already aware of its own conditions and limits. They fail—especially in the assumption of a dialectical orientation—to properly recognize the problem of complexity. The picture theory of consciousness treated the complexity of the world and that of consciousness as symmetrical and was therefore unable to properly investigate these, either as a problem or as an accomplishment. We cannot go back to this way of looking at things. The dynamic-processual concept of consciousness, on the other hand, does offer a basis for functional problem and performance orientation since it includes asymmetry with respect to the "outside world," but it should be thought through again as it relates to the problem of complexity.

An analytical distinction will help us here: we must distinguish conceptually between the immediate moto-sensory data of experience and their co-represented selectivity. Consciousness is not the totality of actually experienced impressions, but rather constitutes itself as their selectivity.[12] It is thus not the introduction of data into the psychic system that consciousness regulates, but their selection capacity; not the input/output processes, but the internal processing of impressions from the environment; not the material of experience but what it does accomplish. Gotthard Günther is thinking along similar lines when he starts with immensely large quantities of possible information and sees in this the basis for the appearance in evolution of interpretive skills made possible by means of selection and redirection that we call consciousness.[13] This also makes it clear that conscious experiencing is meaningful experience processing and can be nothing else.

The clarification of the concept of meaning and its function as the structure of consciousness enables us to make certain distinctions, one of which in particular should be gone into a little more closely. We must distinguish between *meaning* and *information*. Sociological theory is not used to working with a pregnant concept of information, let alone making an explicit distinction between information and meaning. In other disciplines too—in linguistics, cybernetics, information theory—there is a lack of clarity, or at least no uniform opinion, regarding these concepts and the way they are related. However, only a very short sketch should be needed to show that this distinction is of central importance for the construction of a scientific concept of social problems and that in particular the phenomenon of communication cannot be adequately understood without it.

Meaning, as I have tried to show here, overtaxes the potential of actual experience by including and presenting what is not directly experienced. This occurs, however, only within an individual life of consciousness, in a world constituted through pluralistic system formation. This life of consciousness is processlike in nature; the contents that are actualized in perception or thought change ceaselessly from moment to moment. Meaning functions here as a selection rule, and not—or only secondarily, viz. with the help of language—as an actual content arising or appearing in consciousness. In the ongoing flow of experience, reports about the world are constantly crossing the threshold of consciousness, both from the outside and from memory. Such reports take on the character of information by being consciously interpreted—i.e., with the aid of meaning—as selection from other possibilities. The information value here lies in the selectivity of the reported event, something meaning makes possible but does not actualize. Information thus always involves some element of surprise (be it ever so minimal). It rests on a probing of the future by meaningfully structuring expectations, but it is not by fulfilling the prognosis that it informs; it is through the more or less surprising details or deviations in what had been expected.[14]

The difference can be made clear by using a practical criterion. On being repeated, a message or report loses its information value, but not its meaning. Unlike the concept of meaning, that of information is always to be understood relative to an actually given, constantly changing state of knowledge and an individually structured preparedness to process information. The very same meaning complex can thus result in quite different information, depending on when and by whom it is actualized in experience.[15] What one person takes for granted might be surprising to someone else; and the same is true across time: a book that today is hard to understand and full of needlessly complicated sentences may seem quite informative tomorrow, if the reader's expectations have been so restructured that this same book is now able to raise questions or to answer them.

This distinction between meaning and information makes it possible (or rather: necessary) to undertake a reinterpretation of two basic concepts of sociological theory and method—something that amounts to the breaking down of a kind of natural preconception. Experience *(Erfahrung)* and communication will appear in a correspondingly different light. *Experience (Erfahrung)* is surprising information that is structurally relevant and leads to a restructuring of the meaningful premises of experience processing within both concrete and abstract (depending on the circumstances) functional contexts. ("The pliers aren't where they're supposed to be"; "people are undependable.") Experience *(Erfahrung)* is never the pure unmodified arrival of what was expected—when I go downstairs, the fact that the steps are still there is not Experience but merely the informative modification of individual aspects of what was expected. Experience *(Erfahrung)* can, thus, never directly confirm those meaning-bestowing expectations on which its own possibility rests, but can only do so indirectly, by not modifying them. Experience *(Erfahrung)* is an ongoing reconstruction of meaningfully constituted reality brought about by dealing with unfulfilled expectations, by the normalizing processing of information.[16] Experience

32   MEANING AS BASIC CONCEPT

(Erfahrung) is made scientific by increasing its information value, in particular by making its relevance more abstract and by multiplying the number of possibilities it chooses from—and not by the confirmation of existing expectations or opinions.[17]

Communication is not at all what the commonly held view (and quite often the ill-considered scientific use) of this concept takes it to be, viz. a process of "transferring" meaning or information;[18] it is a shared actualization of meaning that is able to inform at least one of the participants.[19] The notion of such a "transfer" already runs into trouble by assuming the identity of what is to be transferred and thus that possession is relinquished when this transfer takes place, i.e., by assuming some form of zero sum. What remains identical in communication, however, is not a transmitted, but a common underlying meaning structure that allows the reciprocal regulation of surprises. That this meaning fundament is itself historical in nature, i.e., that it arises within the history of Experience and communicative processes, is another matter altogether and does not contradict my thesis that communication does not transmit or transfer meaning, but rather requires it as pregiven and as forming a shared background against which informative surprises may be articulated. Nor, of course, does it transfer information, since information has its identity as something occurring at a particular point in time and not as something that endures in time and is able to be transferred. What we have in the case of communication, then, is not the transfer of things but the allotment of surprises.

Communication affords a socialization of surprises and thus also some help in dealing with them. In this sense of a balancing out of information, communication takes place in all situations where one willingly or unwillingly allows another access to the meaning of his experience. Language is not required for this; a slight frown, for example, or a rearranging of certain objects, may be enough: the undercooked potato is pushed to the edge of the plate—and the cook understands! Language is a secondary specialization of the communication process (although it then becomes the basis for all higher evolution of meaning). Language enables communication as such to be differentiated or separated out of the rest of action and thus renders practically unlimited the types of behavior that can be used to inform others. Only thus can meaning be freed from the concrete situation and itself made a content of the processes of consciousness in such a way that meaning can also regulate the selectivity of meaning. Language might then be defined functionally as an increase in the selectivity of the communication process, and communication as an increase in the selectivity of the perceptual process.

Although not all of the details can be anticipated here, these analyses should have far-reaching implications, especially for sociology, which has traditionally devoted very little attention to the problem of information. They help explain why, with the very first beginnings of meaning formation on the basis of organically differentiated systems, an informing effect seems to set in almost automatically, giving rise to life situations in which it is advantageous to more fully realize the possibilities of increased selectivity given here, i.e., to develop language. They point to a fundamental relationship between system structure and communication, viz. that increasing system differentiation also increases the likelihood of differing information situations within the system and thus makes communication process more difficult and demanding—if no institutional countermeasures are taken (the development of communication plans, special languages, delivery systems for mass communication, the separation of informing from motivating information, and so on). Above all, they cast doubt on the wisdom of regarding verbal communication as *the* basic model for all types of action.[20]

Instead of following up on these points here, I want to further develop what has already been said about Experience *(Erfahrung)* and communication. In both Experience and communication, information is normally "normalized," i.e., dealt with by being so interpreted that it accords with already existing or accepted meaning.[21] The unknown is assimilated to the known, the new to the old, the surprising to the familiar. Even if little white mice start popping out of the soup bowl, what we have are still quite ordinary animals, familiar in type, which someone must have put in there— an isolated joke and not a breakdown in the constants of life and nature we have relied upon until now. The surprising or anomalous event is grasped as concretely as possible and is as symbolically isolated as possible, so that the required structural changes can be kept limited in scope and made to proceed along predictable lines. In particular, in social systems with elementary face-to-face interactions, where reaction times are extremely short, such normalization is an almost unavoidable mode of processing information.

This does not mean that information can never be used for a critique of existing structures, but a special effort and special measures within the system are required if this normalization tendency

34 MEANING AS BASIC CONCEPT

is to be changed into a tendency for existing structures to be questioned or problematized and information evaluated as a symptom of impending crisis, as cost, as dysfunction in the prevailing order, or somehow or other looked at as a possible source of alternatives. The prerequisites for this taking place would have to, and can, be investigated using the ideas put forward here. For one thing, they are likely to be found in the security reserves one needs to be able to cope with destabilized structures and a largely open future. In addition, ongoing, chronic structural criticism would appear to require some degree of control over incoming information. Structural critique is hardly possible in the face of little white mice. Suitable information seldom, if ever, appears on its own; it must be specially produced, brought to light by uncovering some latent aspect of the existing order, or retrieved from the existing decision-making process by incongruent questions. It requires specially constructed hypothetical theories, alternative structures, comparative statistics, possible utopias—in any case, special constructs of meaning that let Experience become informative in a way that promotes structural change. All of this is possible only in very complex societies, and even there it is by no means certain that in Experience and communication meaning structure and incident information will come to be related in a way that will stabilize this high level of complexity.

Compared to information, meaning has been looked at here, so to speak, from the outside—as the premise of the ongoing processing of experience. A second approach will have to try to shed some light on how this functional, negating performance of meaning formation affects the form of the contents of experience as they stand before consciousness—how, in other words, meaning appears within experience. I will disregard here the fact that such an appearance, as an entering into the field of actual consciousness, may also possess information value.

The constitution of a world full of not-yet-actualized other possibilities has consequences, first of all, for the form in which possibility itself is experienced. The complexity of other possibilities in practice excludes (or at least renders extremely inexpedient) thematizing all possibilities as subjective possibilities of one's own movement—for example, regarding the table before me as the sequence of impressions I would have if I went around it, observing carefully, and pulled out all of its drawers, lifted it up, checked to see how solid it was, and so on. If the environment were ordered in this way, I would be committed to a fixed sequence of experience, and this would require some basis or support in instinct, or else expose me to continually unfulfilled expectations, to ongoing disappointment.

With a significant increase in the available possibilities of experience and action, however, it becomes difficult to construe all complexity as ordered on the time axis. One would have to either (a) construct an endless chain of ordered experiences, with the unacceptable risks of fixing firmly on this particular sequence and no other; or else (b) make provisions for an incredibly large number of conditioned alternate sequences, but then—lacking the power of prediction—not be able to choose from among them with any continued success. The time dimension must therefore be relieved in its capacity as the sole bearer of complexity. This is achieved by material or objective complexity being presented in consciousness, and this by my actual, given experience indicating the presence of a world full of other possibilities, so to speak, representing the world.[22] The subjective frame of reference that so easily suggests itself—the one that consists in taking my own situation and the accessibility of things for me as the condition of the possibility of possibility—must then be relinquished. Possibility must be objectified, i.e., it must be seen in the things themselves. It must find the condition of its possibility in a world order that does not depend on me and thus at the same time must make it possible for me to determine what sequence my experience is to take, a choice that is free and only "motivated" by the environment. The world must be ordered, so to speak, from no particular perspective, but still in such a way that the choice of the next perspective of my experience does not entail extraordinary difficulties and is even suggested to me; and in practice this means above all: that it can be made without too great a loss of time.

This is accomplished by means of identification. The possible is located in something that remains identical and actually has its identity in just this holding together of Possible and the Not-possible. Meaning then appears as the identity of a complex of possibilities. This, however, furnishes us with only a formal description and not a functional analysis. Identity has long been regarded as one of the attributes of Being and thus as self-explanatory. The central problem we started with, however, will allow us to ask about both the function of Identity and the way this function is fulfilled.

Identical meaning fulfills its function of the constitution and reduction of possibilities of experience and action with the help of negations—not in such a way that the nonidentical is simply and fruitlessly negated, but rather through a *differentiating negation*, in which several mutually independent dimensions of experiencing the world are constituted. The multidimensionality of the world is a precondition for the constitution of identical meaning (and vice versa). These dimensions can be distinguished in experience itself as the *social*, the *temporal*, and the *material* or *object* dimensions.[23] That they may properly be distinguished is shown by the fact that negations within one dimension do not necessarily imply negations in the others.

Meaning appears *materially* or *objectively* in Otherness, in being-one-thing-and-not-another: a horse is not a cow, a number not a pleasure, quickness not a color. Identical meaning stands as well-specified or specifiable complex against a background of indeterminate and negatably negated other possibilities. This requires the components of negation already discussed, generalization and reflexivity. It is only through them that Otherness can be constituted in such a way that it does not exclude the possibility of specifying or of negating negation, i.e., does not exclude the existence of everything else, but actually preserves it and only neutralizes it. The mutual negation found in Otherness thus also entails mutual accessibility and—as a possibility—mutual confirmation.

This Otherness of material or objective meaning, this constitutive relationship between objective identification and negation, is not be be understood on the level of already constituted meaning. I am not referring here to one particular empirically given mundane fact among others, nor to the question of whether (and then why) something that has Being *is* or not (a question that proved too much for an ontologically oriented metaphysics). Identity and negation are rather regulative components of meaning-constituting experience, premises for the processing of experience. This insight —which can also be formulated within a destructive antimetaphysics (e.g., Heidegger's thesis that Being is not the being of what is being) or within analytical philosophy (the thesis that "to be" is not an admissible predicate, i.e., does not constitute a meaningful statement about the real world)—is also relevant for sociology. It lies behind the assumption that there may be social contexts that are not merely given in the referential horizon of already constituted meaning, but instead genetically regulate meaning-constituting experience.[24]

This pattern of negation involved in material or objective Otherness leads to identifications only if it remains restricted to the object dimension. It presupposes that the units of meaning and the Otherness involved in them are not at the same time negated in the social dimension or the temporal dimension. In other words, where meanings or even specific negations (There is no devil!) are concerned, we must be able to count on consensus and duration in time. Material negation must be supported by social and temporal nonnegation; otherwise, the intended object of experience would disappear definitively and even negation could no longer bring it back or make it accessible again.

These analyses make it plausible to regard as the most important variables of the material or object dimension the extent and form of compatibility with other possibilities. Recent psychological research suggests, in addition, that single continuum may suffice in conceptualizing this variable, one that runs from more abstract to more concrete premises of information processing.[25] In the limiting case of being completely trapped within the concrete sphere, experience knows no possibility of negation and is regarded by psychologists as pathological. With increasing abstraction it attains improved chances of encompassing other possibilities of experience and action, as well as more highly differentiated possibilities of negation. All of this (as we shall see at the end of our investigation) can be included in a theory of the evolution of meaning systems that is of interest not only to psychology but to sociology as well.

My argument for the other dimensions has a formally analogous structure. It will require, however, a somewhat sharper sense of abstraction and a firm hold on the insights concerning negation that I have attained thus far if I am to bring out the parallels.

The *social* dimension of experience is constituted in conjunction with material or objective identification by a nonego being recognized as another ego, being experienced as the bearer of its own, albeit different, experiences and perspectives of the world.[26]

Where the experiencing subject finds itself confronted with another ego, it can learn to actualize this other's perspectives, to experience its experience. The other's own perspectives become my own other perspectives; the other's actualization of them guarantees the possibility of their actualization for me as well. The perspectives may be exchanged by exchanging positions; they are transferrable because they are able to find support in the identical meaning (i.e., material or objective non-otherness) of an object and preserve its identity in the course of the exchange. Only thus can selectivity be constituted in consciousness.[27]

38 MEANING AS BASIC CONCEPT

An essential requirement for this process of the intersubjective constitution of a meaningful world of objects is the *nonidentity of the experiencing subjects*.[28] Only this makes possible the separation of the subject living inextricably within his experience from the contents of this experience; his objects are also those of the other subjects and thus have their autonomy in what makes them accessible for all—in their meaning. This leads to a repair of the perspective-given distortion of the world and this in turn to a reflexive consciousness of one's own perspective as only one among many that are possible. As such, it can then be consciously chosen from among those that other coexperiencing subjects hold ready in their experience. The other subjects involved here relieve the individual's actual consciousness of having to make possible, alone and solely on the basis of the contents immediately given to it at any time, all of the possibilities of the world, i.e., relieve it of having to function by itself as the condition of all possibility. Only thus can a complex world be constituted as the horizon of the potentialities of actual consciousness, as the unmanageable and overburdening source of all selection.

This helps explain why communicative relationships between subjects help to determine what is possible as a world for these subjects. The boundaries of communication (a much-discussed topic), including both the structure of language in general and the peculiarities of the particular language being used, tend to restrict what can be articulated as meaning. But that is not all. Even the question of who is experienced as another experiencing subject is answered differently in different societies: only in highly developed societies are all persons (and only persons) included.[29] The degree of concreteness or abstractness with which the other subjects function also varies greatly from society to society, as well as over time in the general evolution of society, and even among the various provinces of meaning within a society, for example, in the family[30] or in science.[31] The crystallization of the other person as a "subject" (that means: as a consciousness standing behind the world) out of his typically known characteristics and situations, out of the context of living together and proper behavior out of role and status structures: in short, the reduction of the relevance of coexperiencing to mere sense perception and the ability to use abstract concepts—this requires feats of abstraction which come to be institutionalized only very late and only in certain areas of social life. Unlike the I-ness attending lived experience, being-a-subject is not something that is "innately" given or that phenomenological reduction can bring to evidence, but instead a form of human self-constitution which appears very late and only where the necessary societal conditions have been realized.

Although these analyses point the way to a sociology of knowledge, they cannot be taken any further here, since we have to remain on the level of the general constitution of meaning and complete its investigation. This still requires consideration of the temporal dimension—something, incidentally, that should help us understand how the abstraction of man to subject is accompanied by changes in his notion of time.

The features of the *temporal* dimension of experience necessary to make meaningful identities possible have a very complicated structure, both considered in themselves and as they relate to the other dimensions of experiencing the world. I will start by making a distinction between the temporalness and temporal locus of the constituting experience, on the one hand, and the temporalness and temporal locus of the constituted meaning, on the other. It is absolutely necessary that this difference itself be experienced, if the horizon of meaningful experience is to have temporal extension, i.e., if one is to be at all able, in the present, to imagine past or future meaning (or present meaning extending into the past or the future). Such temporal extension of meaning that is nevertheless actually presently experienced functions as a sustaining negation and, in this way, makes it possible to use the temporal dimension itself as a representation of complexity: the steps necessary to realize a future goal can be chosen in the present.

The difference itself must be brought forth and borne in actual, ongoing experience; it must itself be part of what is presented to consciousness. This is why the analysis of the temporalness of the ongoing present of lived experience takes precedence over the analysis of constituted meaning; it cannot be seen in the temporal structures of the meaning itself, in its permanence or impermanence, its quality as something occurring or something enduring. Above all, the foundational relationships between the temporal dimension and the social dimension are to be found on the level of constituting experience, i.e., always in the actual present, and not in the sociality of the constituted meaning (not, for example, in the question of who a certain thing belongs to or who made a certain statement).[32]

On this level of meaning-constituting experience, the social conditions for the constitution of material or objective meaning involve an important reduction of the possibilities of the temporal dimension: there can be no time difference between experiencing subjects. Their ongoing, actual experience must be synchronized, i.e., must, according to their own understanding, take place simultaneously. Not only the Present itself but also its temporal horizons of Past or Future must be equalized and—although this clearly contradicts our direct Experience of the very different "closeness" past and future may have—must be located at the same distance. For this simultaneity in temporal experience to be maintained, the tempo must also be brought—and here too in spite of our direct Experience—to a uniform flow; i.e., variations in the flow of time must be attributed to purely subjective perspectives and thus considered illusory. "In itself" this is not so obvious: why should the other's meaningful look not be my future? Why should his annoying reminders not simply be defined away as part of my past? These possibilities, however, are excluded, because such an individual meandering about in time cannot be kept up without overtaxing the social and material/objective dimensions with complexity and the need for negations. The myriad possibilities contained in the multiplication of temporal with social complexity can be more effectively dealt with by means of synchronization. No one may jump ahead into an other's future or remain trapped in his past:[33] we all get older together and at the same rate.[34] This means that, along with the unitary character of intersubjectively constituted time, the transferability or exchangeability of perspectives of experience is also guaranteed and, along with this, common access to the world. Only in this way are we assured that all possibilities lie in the future and none in the past—that they remain possible for everyone. Only in this way can an other's actual experience serve as guarantee for my own potential experience—of course, with the reservation that actualizing it takes time and presupposes that I have some control over my future. Time difference for different experiencing subjects is now seen only in standpoint-dependent variations in the use of time in attending to possible experience or action.

Only with intersubjectively synchronized time does it become possible to fix meaningful identities in their very own temporal reference schemes—to date them, assign them to the past or the future, mark out the temporal limits of existence, applicability, perceptibility, and so on—without this limiting our ability to consider or thematize the meaning itself. In everyday life, therefore, it is quite sufficient to orient ourselves within already constituted reference schemes—on things and events and movements in time.

MEANING AS BASIC CONCEPT 41

However, any sociology that seeks to clarify the social bases of meaning-constituting experience will have to look behind these ready-made conceptions. Two examples here should at least help to indicate the necessity of such recourse to meaning-constituting experience, one concerning the perception of time, and one dealing with the problem of security.

As with the self-determination of the subject, we find that the subject's perception of time also involves feats of abstraction that are closely tied to the given social structure, i.e., that are evolutionary variables. Time is so constituted that it affords room for the possibilities of experience and action that are made possible by language and social structure. One mode of abstraction follows from a loosening of the interdependence among the various dimensions of experience, with their increasing separation doing away with old ties and limitations. Time is abstracted out of the objective/meaningfully constituted world; it loses its intrinsic relationship to the familiar flow of things and events, its ties to astronomical or life rhythms, to feasts and the yearly cycles; and it loses its ability to mark out certain time points as such, as *kairos*, even its ability to act as a cause—e.g., in the sense that the simple duration of life *makes* one old. It becomes an abstract continuum of time points along which everything can move in accordance with "laws" or "systems" which are not themselves time but are only measured in time.[35] Parallel to this, the ongoing lived present—the standpoint of intersubjectively communicable experience—recedes from its position of dominance over time consciousness, allowing the orientation of the present to be displaced from the past into the future.[36] It is no longer history with its already reduced complexity, its already excluded other possibilities, that has absolute primacy over the present; it is the future. The past is now finished, can now be regarded as over and done with; it no longer intrudes into the present in the coexperience of the dead or the continuity of guilt. It remains important as the structure of systems, as capital, in the sense of an accumulation of money or knowledge or power, or as history, in the sense of something that can be uncovered by future research—in any case now seen as something securing the freedom of future dispositions.[37] As human freedom increases, the obligation of tradition comes to be replaced by the unavoidability of selection. "We were," Sahlins[38] comments on this development, "*chosen* people; now we are *choosing* people."

To the extent that time leaves the future open and all possibilities possible, it is the ongoing, immediate present of actual lived experience that becomes problematic. That which alone endures is now reinterpreted as a minimal moment moving toward the future along a fixed scale of well-labeled time points. The future is, thus, no longer that which is moving toward us, but instead that open horizon into which, carefully choosing our direction, we are moving. The lived present is no longer that which endures and on which time flows by, but just the opposite—that which is moving in time. The present must now hold in store some not fully determinate elements, which can be filled in only by future dispositions, or else fully determinate elements made so that they can accommodate future reinterpretations.[39] And along with this, we find that materially/objectively related elements of meaning, which are constituted together in the common present of lived experience, come to be desubstantialized, instrumentalized, and, finally, functionalized, and can now have meaning only in this way; that it is in its causality—and not as the intrusion and realization of Being in the Present—that human action is understood and rationalized.

For Max Weber, who believed it was possible to use an ends/means scheme to understand the intended meaning of human action in its rationality, and to find precisely here a guarantee of intersubjectively communicable (scientific-sociological) statements about human action[40]—for Max Weber this highly derivative and presupposition-laden conception of action still remained the unquestioned theoretical basis of sociology as well. Since then doubts have increased.[41] The place that, for Weber, was so firmly occupied by value rationality must be investigated somewhat more carefully. And this will reveal the unrelinquishable presence of meaning-constituting experience, whose fulfillment is prepared by the future and assured by the past but can take place only in the present.[42] Only in the enduring present of meaning-constituting experience can security be offered, fear held in check, and trust extended.[43]

A sociology that does not simply treat meaning as a cultural artifact, but also wants its concept of meaning to encompass meaning-constituting experience in its social dimension and its relationship to evolutionary and social-structural factors, will be forced to develop new and difficult ideas about time. Its concept will have to encompass a twofold possibility of thematizing time. On the one hand, constituted meaning can be referred—as an event or as something objectively given—to objectively fixed time, through which subjective experience progresses, continually transforming its future into its past. According to this way of seeing things, the qualification as future or as past is purely subjective: what is proper to time itself is only its irreversibility. On the other hand, we can imagine the enduring present of meaning-constituting experience only if this itself is regarded as fixed and the meaningfully constituted events as in flux. Both versions of time are possible and both are legitimate (which confirms the old insight that the "nature of time" cannot be expressed in the opposition between what is fixed and what is flowing). Time itself—if we follow these insights—can be only the possibility of these two quite contrary conceptions, only what makes the contradiction possible.

To examine this further, I will have to return to my thesis of the functional primacy of negation in meaning-constituting experience. All identity is constituted by way of negation. In the horizon of time, identity can be fixed on well-labeled and fixed time points or on stretches of time, and then appears as an event or a sequence of events. This identification has its principle in the variability (i.e., in the nonidentity) of the situation or consciousness with respect to time; it is in the face of changes in qualification as future, present, or past and in the distance that this involves that it remains identical. However, identity can also be referred to the enduring nature of consciousness with its fixed horizons of future and past, and it is then with respect to the passage of time points out of the future into the past that it remains indifferent. It then rests in the eternal presentness of consciousness and finds its unity in the negation of the relevance of changing time point–relative localizations—and it is just this that is meant here by eternity. Which interpretation of time is chosen—the ancient or the modern—depends on which possibilities of grasping and reducing complexity are demanded in human experience, i.e., depends on what state of human existence has been reached. The choice of identification and negation made within the temporal dimension also determines what ensuing problems will have to be resolved. The modern subjectivization and mobilization of consciousness has an impact on time, makes it difficult to maintain enduring states and makes time a scarce resource. What time itself is cannot be expressed adequately in either of these opposing interpretations and remains hidden beyond the reach of sociologists in the identity of the nonidentical.

The ideas presented here can be summarized as follows: meaning is the form for the ordering of human experience, the form for the intake of information and conscious processing of experience, and it makes possible a conscious grasp and reduction of high complexity. A closer analysis must go back to the meaning-forming achievements of experience and leads us to a complicated network of negations with whose help identities are constituted within a multidimensional, socially, temporally, and materially or objectively complex world. This gives rise to the impression of an objective world that is already limited in its possibilities, a world that is not dependent on the actual course of experience at any time but can, instead, be thought of as the domain from which all such experience is drawn. All meaning points to this world in its entirety and all meaning provides access to it. This meaningful construction of the world as the reference horizon of consciousness involves high risks, for man lives on the basis of a physical and organic system under real conditions which he interprets as world but cannot change at will—which he constitutes as meaningful-identifiable but does not create. He accepts, in other words, the risk of negation. His meaning structures remain susceptible to disappointment, to nonfulfillment. His world is contingent; it could be otherwise. This means that there is not only the programmable problem of selection out of an excess of other possibilities to be considered, but also the risks that selection involves; not only rationality but fear as well.

Before I can analyze the structures and processes involved in this problem of overcoming fear and dealing with unfulfilled expectations, I will have to try to arrive at an adequate understanding of the nature of the problem—and for this too we will have to turn back to the foundations in meaning-constituting experience. Our look at the temporal dimension has already made us aware that final certainty can be attained only in the present: only what is immediately present is evident, obtains fully, and admits no other possibility. The most that can be done for the future is to secure, in the present, some certainty equivalents—money, for example, or the well-bred confidence of never committing a *faux pas*. In the material or objective dimension, contingency demands an ability to learn, i.e., an ability to adaptively or innovatively alter the structures regulating information processing. Such learning ability appears to rest in the relationship of abstract to concrete premises of experience processing and to increase the increasing abstractness of the system structure, for both psychic and social systems.[44] In its dependence on these conditions, it can then be investigated by sociology, especially a theory of the evolution of society. But what must be of particular interest to sociology is how contingency appears within the social dimension. Here it assumes the form of *double contingency*, something we will have to look at more closely, since this formulation of the problem determines what social structure can be.

When I speak of double contingency, I have in mind that element of dependency that guides mere possibility on its way to being-so-and-not-otherwise. All experience or action that is oriented to others is doubly contingent in that it does not depend solely on me, but also on the Other, who I must regard as an alter ego, i.e., as just as free and unpredictable as I am.[45] The expectations that I address to an Other will be fulfilled only if *he and I* both do what is necessary for this—and this condition is itself reflected upon and forms part of our expectations. All of this involves an involuted risk of failure, which is increased even more by our being conscious of it, and, at the same time, involves an indication of the direction the solution to the problem must take: one must, under these conditions, be able to have expectations not only regarding others' behavior but regarding their expectations as well—for only in this way can the regulative principle of the other's freedom be incorporated into my expectation structure.

Social structures do not take the form of expectations about behavior (let alone consist of concrete ways of behaving), but rather take the form of expectations about expectations.[46] In any case, it is only on this level of reflexive expectation that they can be integrated and maintained. The sociality of meaning, for example, the social aspect of the meaning of some act, is not exhausted by referring to the fact that another person (of a certain general type, with particular individual characteristics, a personal history, etc.) *exists;* it lies instead in the fact that the *intended* meaning can be recognized, and this recognizability has structural relevance, for it tells us something about what the other *expects*. With the help of experience processing based in meaning, someone having or forming expectations can also take into account the expectations directed at him, can even expect that the other harbors expectations about his expectations and that the security of these expectations must not be upset (or: can be destroyed).[47] Expectations about expectations save on communication and, above all, on conflict-laden confrontations in real tests of opinions.[48] Only in this way is it at all possible to deal effectively with the great complexity of social expectation networks and their double contingency, given only a very limited potential for direct and conscious attention. The expectability of expectations is an indispensable requisite of all social interaction that is guided by meaning. It is prior to the

46 MEANING AS BASIC CONCEPT

thoroughly secondary distinction between conflict and cooperation, since both of these types of interaction are possible only where expectations can be expected.[49]

This brings us to a second version of the problem of double contingency, to the question of how it is possible, without actually partaking of an other's consciousness, to successfully expect others' expectations. This can be assured only to a very limited extent by concrete familiarity with certain persons and their unchanging characteristics, personal histories, and habits, and only in certain respects by explicit communication. For this reason, meaning itself must contain, as it were, syntheses of expectation on which one may base one's behavior without this involving unacceptable risks. This occurs to a large degree in the form of a *right to make assumptions*. Types and rules of meaning (for example, the type "question and answer" or the rule "Sundays from 11:00 to 1:00 is visiting time") are developed for use in social interaction. These allow for the assumption of a corresponding set of expectations (e.g., that the questioner expects an answer) without this first having to be checked or an overt understanding having to be reached beforehand, and they *protect us in cases of error or disagreement:* Whoever is not in agreement must as least voice this explicitly and take on the responsibility for correcting expectations; it is he who must carry the burden of initiative and of argumentation—and in many cases of norm-violating, immoral expectations and behavior. The "blame" for the discrepancy is given to him and not to the person with the expectations usually appropriate to the meaning in question. Provisions for cases of disappointment are thus built into meaning itself, provisions that make it possible to be assured in advance that even in cases of nonfulfillment of expectations or expectation-expectations there will still be a workable basis for behavior.

I am now in a position to show how *normativity* and *technicalness* [lit.: technicalizability. Tr.] are rooted in the meaningfulness of human experience, are first rendered possible by meaning, but then needed to be developed in a certain direction. Meaning becomes *normative* to the extent that what is provided for in cases of disappointment or non-fulfillment of expectations is a continued maintenance of these same expectations, i.e., to the extent that learning is excluded.[50] Norms are contrafactually stabilized expectations, which are protected—at both the level of behavioral expectations and of expectation-expectations—against the symbolic, discrediting implications of nonfulfillment. Meaning becomes *technical* to the extent that the process of experience is freed from its accompanying meaningful references—relieved, so to speak, from having to include the entire world—and can then go through an abstractly specified sequence of selection steps (e.g., a mathematical calculation, or the step-by-step composition of a work of art, or a sequence of choices of means appropriate to a particular goal) without thereby being irritated or jeopardized by that neglected horizon of other possibilities. Thus, the normative and the technical ultimately have the condition of their possibility in meaning-constituting experience and can be properly investigated—if this is kept in mind—only by considering the evolutionary and social-structural conditions of their development: it is only complex, highly differentiated societal systems that make possible the establishment of improbable, almost arbitrary norms[51] and the improbable, technical specification of social contact.

As we turn back to the conditions and forms of meaning-constituting experience, we can see that the complexity and the contingency of other possibilities of experience and action are related, and that we might also look for some relationship between rationalization and the way fear is dealt with. System evolution can lead—especially on the level of the overall society—to a considerable, ultimately incalculable expansion of the realm of possible experience and action—but only if the forms of rational selection become correspondingly more effective *and* the absorption of uncertainty and fear is still possible in the face of high complexity, without this impeding rationality. Sociological theory must be capable of understanding this relationship. Although such topics as the secularization and rationalization of modern society, system differentiation and solidarity, society and community once occupied classical sociologists, they are in danger of being lost in more recent theoretical developments. This discussion would have to be freed from perennially fruitless dichotomies and taken up again within a more abstractly conceived theoretical framework. In doing so, we must recognize that the problems of complexity and contingency present themselves in several dimensions and, thus, require relatively complicated institutional solutions. We may assume that the relationship of concrete to abstract premises of experience processing (focused, for example, in the relationship between family and workplace) must be appropriate to the forms in which future is represented and to the current solutions for the problem of double contingency by means of expectation syntheses—that we have here, in other words, problems of compatibility and that not all combinations of meaning are possible.

48 MEANING AS BASIC CONCEPT

Such relationships can be seen if we look at the civilizations of the past. We see that a rational mundane praxis capable of transcending the immediate present became possible only with the individualization of fear and the means of overcoming it, and with the abstraction of the relationship between morality and religion. However, the institutional solutions found for these problems were rendered obsolete by later developments of society. Modern requirements for rational selection out of very high complexity can be met only by means of disposition over structures, something that makes us conscious of their contingency (positivity). In what form the risks that this entails may be made bearable and how they appear in experienceable meaning is still hard to predict. The beginnings of one solution to the problem can be seen in a clear and pervasive increase in indifference and trivialization; they can also be seen in a strong diffusion of the uncertainties relevant in any given case. But the ability to bear the primary risks of our society—in particular those of the centralization of the political decision-making process and of scientific research—may rest in the fact that for anthropological reasons we are simply not capable of enough fear and that these risks, therefore, cannot be at all adequately indicated in experienceable meaning.

The function of meaning is the indication of, and control of access to, other possibilities. In its horizon of possibilities, in the complexity and the contingency of what it positively or negatively points to, there also lies that variable governing the form in which meaning appears. This form of meaning, which I have referred to as identity and as located in the various dimensions of experiencing the world, is thus itself to be regarded as a variable—as a variable, if I can say this, of transcendental evolution. Keeping this in mind, I must now attempt to describe somewhat more closely just how such an achievement is possible. In doing so, I shall make use of the distinction between *form* and *content* as well as that between *structure* and *process*.

If we take the classical distinction between *form* and *content*, something intended to signify two dissimilar but mutually dependent aspects of how an object may be given, and apply it to our functional concept of meaning, we can also recognize its function. As Heinrich Gomperz has already noted,[52] the form/content distinction allows for a progressive or stepwise approach: in the form one can recognize possible meaning—a conjecture of the understanding, says Gomperz—and can then set about trying to transform this possibility into actual experience. Unlike simple stimuli, form is a perceivable and instructive preview or anticipation of the totality of possible contents. The grammatically correct form of a sentence, for example, gives notice of a meaningful statement and, as we now know, of limited possibilities of variation. It suggests to us that concerning ourselves with this sentence will be worthwhile, but at the same time offers no guarantee that such an undertaking will in fact be fruitful. And it does not exclude the possibility that the form will turn out to be deceptive, the promise empty. Like all meaningful abridgements of experience, this too entails risks.

Such a progression in experience is always necessary where the complexity of what is indicated exceeds what can be grasped directly. The rationality of the form is measured on the basis of the success with which it fulfills its presorting and directing functions and absorbs the accompanying risks. This should help us appreciate the important advantages of verbal and grammatical forms: they allow contents of greater complexity to be grasped by way of the form and are, in this sense, more rational than material- or thing-forms. Linguists also refer to this as the "meaning transparency" of verbal signs.[53] All of this shows us that form is functional —not, however, that it is necessary. It is theoretically possible that the functions mentioned here can be fulfilled, in whole or in part, by some other means. In particular, we may expect that in highly complex, information-rich societies all form-experience would have to be preceded by still simpler, more powerful, and less instructive stimuli which serve to capture the attention—for example, movement, novelty, absurdities, scandals, pain or pain surrogates (loss of money), the higher status of the communication source, and so on.[54]

If we accept this interpretation, form and content are guides or instructions, appearing within meaning itself, for a progressive grasping of meaning. This distinction thus fulfills a function with respect to our basic problem here, that of meaning overtaxing the capacity for conscious experience processing. The same is true in another sense for the distinction between *structure* and *process*. It refers to the necessity, in the actual course of experiencing, of positing some meaning and using it as a regulative premise for directing our experience—this too as a means of relieving consciousness.

Meaning is sometimes even defined using this distinction—for example, as operation in accordance with given standards, as the application of a code, as speaking a language. We should note here, first of all, that meaning does not consist simply of the rules in themselves, or of the ideal existence of abstract entities, but is constituted only in the practical application of such rules, only in the performance or carrying out of the actual life of consciousness. Meaning is the functioning of premises. The structure/process difference is nothing other than the way the ongoing, actual experience of meaning functions to free itself from having to think through all of its possibilities at the same time. The currently unactualized possibilities function as premises of the ongoing processing of experience, and this in a form that is "instructive" for actual experience. Such instructiveness and such proximity in experience of what is still removed from it can be achieved in very different ways —not only through word formation or categorial abstraction, but also, for example, by the way space is represented (I know that New York is far away and that at the moment it's not possible to speak to someone there; I would have to travel there or try to telephone).

What has been said here offers us an opportunity to reconsider the *relationship between meaning and language* and correct the widespread overestimation of the role of language. This is not to deny that language plays an indispensible part in the meaningful constitution of the world. Without language, the only perspectives of experience that could be objectified would be extremely simple, extremely poor in references to other possibilities. Without language, we could not actually intend negations and probably not even experience them explicitly—a loss, as we have seen, of an essential aspect of the constitution of meaning. And last but not least, we may assume that without language the ordering of experience through a consciously functioning structure/process difference could never have been developed and could no longer be learned fast enough—all of this not to speak of the impossibility of establishing and elaborating a cultural tradition without language. But one may nevertheless have reservations about the attempt to base sociology and its concept of meaning on a theory of language. First, it is not at all clear how a theory of language could bring adequate clarity to the concept of meaning, since it already presupposes this notion in all of its basic concepts with the possibility of their identification. Also, the central problem that my concept of meaning refers back to and for which I have chosen the formula *contingency and complexity*, while indeed being a problem that can be talked about, is not a problem of language alone. And second, language itself is nowhere near capable of producing the clarity and certainty of meaning necessary in practical life. The possibility language offers of lying should be enough to convince us of this. Language serves primarily to keep possibilities open, to allow access to unforeseen meaning combinations. Although language does determine the conditions of possibility, i.e., the conditions of linguistically possible sentences, it does so within an extremely broadly conceived, universally adequate framework which allows far too much and, therefore, requires further selection mechanisms. Language alone is incapable of establishing meaning: this requires, in addition, systems whose particular structures define narrower conditions of possibility, i.e., define additional boundaries within the domain of the linguistically possible. For the realm of meaningful experience and action these are psychic and social systems of the most diverse kinds.

There is little difficulty in giving some plausibility to the function and the necessity of such boundary-establishing and boundary-maintaining systems within the sphere of communication opened up by language. But just what "boundary" should mean in this context is anything but obvious and has never been adequately clarified. The usual explanation on the basis of the difference between inside and outside or system and environment does not help us further here, since it merely offers a reformulation of the problem. The supposed clarity in the notion of boundary has its origin in the realm of physical and organic systems and has been somewhat too hastily assumed for meaning systems.[55] For the realm where experience is ordered on the basis of meaning, however, the phenomenon of the meaningful boundaries of social systems[56] is something that requires a careful clarification, which could start with my analysis of the concept of meaning, especially the discovery of the functional primacy of negation in meaning-constituting experience.

The picture so easily suggested to us by physical boundaries is misleading here in the sense that it suggests point-for-point correlations across the boundaries: where the house stops the yard begins. The boundaries here order one relationship of distance or closeness to another, with these always thought of as something quite determinate. On the other hand, meaning boundaries (and physical boundaries can, of course, symbolize meaning boundaries) order a complexity difference. They separate system and environment as possibility domains of unequal complexity. The environment always has a higher complexity than the system, and ultimately the indeterminate complexity of the world itself. Meaning boundaries mark out this difference and make it available in the orientation of experience. They tell us that well-specified and known (or at least easily recognizable) conditions of possibility of action obtain within the system; outside the system, however, obtain less specified others. If experience is directed beyond the boundaries of the system, the first thing one must do then is to make sure about what system one has now entered. The system "university department" contains no rules specifying whether it is *family, theater, church, science, nightclub*, or whatever—each with its own special structure—that will next capture experience and action. From inside the system, what is "beyond" remains unspecified. The system boundaries are in this sense "open" to the outside and, just for this reason, must also be "open" in a second (more conventional) sense —i.e., they must let through information concerning the environment, which still remains unspecified or indeterminate from the point of view of the system. They have a warning function here: they require us to consider "what next?" and to look around for the next means of orienting ourselves. And they usually also involve some mechanism of consensus, such as those seen in greetings and departure ceremonies, in the avoidance of certain topics, and so on.

Under these circumstances, the problem of arriving at quick and reliable agreement about system boundaries, i.e., the problem of the social regulation of the recognition of systems, becomes more and more difficult as system differentiation increases.[57] Where agreement about relatively complicated matters must be reached quickly, perceptual processes probably play a relatively large role, since it is only in perception that a high tempo and complexity can be combined. That, however, means that system boundaries have to be fairly concretely fixed, that they can be signaled, at least in part, by perceivable things; that buildings, territorial boundaries, persons, gestures, and so on serve to activate the relevance of a given set of rules. Of course, verbal signals are also gestures in this sense and do have boundary-indicating functions, but language by itself is not specific enough: simply saying "mail" to a passerby is not enough to get him to send on my letter. However, such concretely established system boundaries are also dysfunctional in their own way, since their mode of separating can interfere with the functional-structural specialization of the system or burden it with some other subsequent problems. Consider, for example, the damage that can be done to the reputation of a system by the outside behavior of its members, or the unpredictable results of associating the identity or prestige of political systems with fixed territorial boundaries.

Even more than the concreteness of system boundaries, however, it is their "openness" that presents a problem for the system. Where there is a high degree of arbitrariness in what a system must admit as environment, the stability of the system structures can be secured only with the help of symbolic neutralizations. Events in the environment must not automatically be relevant for the system, not even when they involve a consciousness that retains its identity on reentering the system from the outside. Such a return must be possible in spite of incompatible outside experience: it is a sign of stabilizing system boundaries if a marriage does not suffer when the wife discovers that her husband did not spend all of his time at the conference thinking of her; or if the board of directors is not disrupted if some of its members run into one another at a porno film. System boundaries may vary in the sharpness of their definition and systems, in the degree of their exclusivity. Systems that demand a particular "conviction" on the part of their members have more difficulties here than those requiring only certain recognizable kinds of behavior. Still, no system can totally do without symbolic immunization—for that would mean doing without boundaries and merging with the world.

All of this is only a reformulation of what I have already established regarding the functional primacy of negation in meaning-constituting experience. Establishing and maintaining boundaries requires both functional components of negation: generalization, in the sense of a general or wholesale turning away from the currently neutralized other possibilities; and reflexivity, in the sense that these negated possibilities remain available and can be turned to in a negation of the negation. In the case of system boundaries, this combination of components has the particular function of screening off domains of meaning that have lower, already structurally reduced complexity from the implication of the world in its entirety that is ultimately carried in all meaning. Well-defined systems make it possible, in other words, to live in the face of an extremely complex and contingent world and yet always have to choose from among only a small number of consciously controllable possibilities of behavior.

Sociology today is generally and almost unquestioningly understood as a science of social action, and this common understanding is shared by both action-theoretical and systems-theoretical positions. Contrasting them requires the assumption that sociology is concerned exclusively with human action, and it is only on this basis that we have controversy about whether it is the concept of action by itself or only that of action system that is theoretically fruitful; or whether, as for example Parsons assumes, the concept of action already implies that of action system,[58] so that a systems theory can be deduced from the concept of action.

If it starts with such assumptions, sociology treats meaning right from the outset as the meaning of acts. It hopes in this way to distinguish itself from other sciences of meaning and to have nothing to do with, for example, the meaning of things or the meaning of symbols conceived without reference to action. However, if we start with the analysis of meaning presented here, these premises become somewhat problematic. One can then ask whether such a delimitation of sociology can succeed, and further, whether the double approach of action theory and systems theory is not ultimately rooted in these premises and therefore undecidable on their basis.

To start with this concept of meaning warns us against interpreting the prevailing view all too narrowly, as if sociology dealt only with a particular portion of the world called "action." The meaning of acts also always ultimately implies the world in its entirety. And snow, too, and personal property, justice, dishes, capitalism, and so on, without themselves being action, can all become relevant within the context of action. The concept of meaning does not —like some set of ontically given properties—mark off those objects that may be dealt with sociologically, but rather regulates the way in which they are dealt with. It does not in itself have the effect of excluding anything but merely requires that we start with the intended meaning of an act when considering what in any given case is significant. But why? And what in this context do I mean by an act? And in what does its identity rest?

After examining how the concept of meaning relates back to the problem of complexity, we can try to find an answer to these questions. In doing so, I shall make use of the distinction between *experience (Erleben)* and *action (Handeln)*. There are, namely, two ways in which the meaningful reduction of complexity can be attributed: either to the world itself or to certain systems in the world. Either the reduction is treated as something given, or else it is brought about by some particular system. In the first case I shall speak of *experience;* in the second, of *action*. Both are processes that take place in systems; both processes require living, behaving organisms that are capable of using meaning in ordering their relationship to the environment. The difference between experience and action can thus not be construed on the basis of the distinction between inside and outside or that between active and passive; experiencing also is living *(auch Erleben ist Leben)* and also involves ceaseless bodily motion. The point of difference cannot be grasped on the level of the organic substratum, or in some visible aspect of man, but lies instead in the formation of meaning itself—namely in the question of how the reduction of complexity is attributed, of where, so to speak, the meaning is "located." Experienced meaning *(erlebter Sinn)* is grasped and processed as having been reduced outside the system; the meaning of action *(Handlungssinn)*, on the other hand, is grasped as having been accomplished by the system itself.

Unlike ontic distinctions, that between experience and action is to be understood relative to some system and, thus, becomes unambiguous only when a system referent is specified. The action(s) of one system can be *experienced* by another. This means that distinguishing between experience and action requires a control level on which the corresponding attributions can be experienced *(erlebt)* and acted upon *(behandelt)*. And this too is given in the meaningful constitution of the world: meaning is experienced as constituted through experience or through action and dealt with *(behandelt)* differently in each case. Furthermore, like all cases of attribution, the distinction here involves an element of convention. The attribution itself can be regarded as contingent, seen as something that could be different, and this contingence for its part can then be problematized. Sociologists should, for example, be interested in the question of how it is possible in simple social systems with elementary face-to-face interactions to reach adequately quick and reliable agreement about what, in any given case, is experience and what is action. One can imagine the problems that would occur in a marriage, for example, if one partner always regarded as the other's action what for that other was only experience, when what was meant as an objective account of external events is chalked up as guilt. Further questions arise if this problem is transposed to the level of the encompassing system of society. One can assume that the line between experience and action is drawn differently in different societal systems in the course of evolution. As system complexity increases, so too can that group of selections that are experienced as action and not as experience, since their selectivity can then be controlled in systems. The province of meaning associated with law, for example, exhibits just such a transformation from meaning structures that can be experienced only—and thus talked about in terms of truth or falsity—into action-based, attributable positivity. In other areas too we find that the experienced action of others is pushing back common experience. Perhaps the reason can be found here for the rise in modern times of a new kind of action concept standing in explicit contrast to nature.[59]

The convertibility of experience into action and vice versa does, of course, have limits, at least limits of practicability. It would be possible, but extremely impractical, to construe the meaning of physical objects as sequences of action on the part of whoever deals with those objects.[60] In other words, there are topics for which it remains of constant advantage over a long period to deal with the processing of meaning as experience and not as action *(Handeln)*. But even these advantages are tied to the construction of psychic and social meaning systems and vary with their structure. We can thus assume, in principle, a functional equivalence between these two reduction forms, experience and action—one, however, that can seldom be put into practice, i.e., can seldom lead to the substitution of experience for action or action for experience.

Although such questions can only be pointed to here and not worked out in more detail, the theoretical assumptions underlying them will have to concern us further. The prevailing use of *action* as a basic category cuts us off from questions like these wherever, in order to distinguish among the various sciences, it gives primacy to the concept of action and sees experience only in the subsidiary function of preparing for or motivating action. This results in parallel claims to founding a universal sociological theory being put forward by the theory of action or action systems, on the one hand, and by the sociology of knowledge, on the other, with no decision on this conflict possible within the framework of sociological theory. This dilemma could be avoided if we turned in our search for basic categories back to the concept of meaning and started with it to derive experience and action as equally ranked, functionally equivalent, but nevertheless different kinds of reduction. This would also make possible a critical examination and justification of the specifically sociological approach in research, i.e., of the question of why sociology treats its particular object, social systems, as systems of action.

My approach readily suggests the answer: since the concept of action gives expression to the reduction of complexity attributable to the system, the system is identified as an action system. Grasping a system as an action system means defining it by means of its own performance, whereby we do not use the classical language of means/ends rationality in describing this performance, but that of selectivity. The choice of the action concept thus follows from the wish for a functional, performance-oriented system definition.[61] Systems theory offers the encompassing conceptionalization and thus leads to the choice of the action concept. This is, not least of all, a result of the switch to functionalistic concept formation, an ontological way of thinking would already be pressed to regard the concept of action as basic simply because systems "consist of" actions.

But what is meant by "consisting of" actions? We will have to rescue this formulation from the traps of language and from a traditional way of thinking that remains caught in its snares, and try to formulate it more carefully. If we regard actions as identified through meaning and conceive of meaning as a complexity-preserving and complexity-reducing reference center, then both a substantial and a relational interpretation of such "consisting" become impossible. Systems are not made up of actions in the sense that these are already there for us, like previously existing objects with their own particular attributes, and cannot simply be taken and placed in relationship to one another. Instead, it is first in systems that the meaning and identity of individual acts are constituted.[62] Unlike experiences *(Erlebnisse),* which derive their identity from that of the intended objects, i.e., from reference to an already reduced order, actions find their identity only within the functional context of systems, through the choice of one or another of the possibilities permitted by the system. Only through a demarcation of its selection achievement does the unity of an action become visible as a slice out of the continuous flow of behavior, which is always choosing from a different constellation of suitably tailored possibilities: at one moment I am reaching for a pen (and not a pencil); then putting this thought (and not that one) to paper; then reaching for the telephone and calling a taxi in order to go out, in view of the fact that my own car is in the garage for repairs, and so on. The identity of an individual act, then, is its current reduction achievement within the reference system which first makes this performance possible by supplying a structure and a history and guaranteeing complementary performances (temporal, social, material/objective) on the part of others.

This highly abstract argument applies to psychic systems, i.e., to premises of experience processing that are identified as personalities, just as it does to social systems. It would be a fundamental error to insist on distinguishing between these two types of systems on the basis of the difference between experience and action—i.e., to define psychic systems as those of experience and social systems as those of action. Contrasting them in this fashion would separate the two types of systems far too sharply and again come close to misinterpreting the difference between them as an ontic-substantial one—while actually the relative "substratum" and the meaning components of experience and action, from which psychic and social systems are recruited, are pretty much identical. According to the present state of theory formation in psychology, social psychology, and sociology, an ontological separation of the proper subject matter of these disciplines is no longer really possible. Rather, it is necessary to take as our starting point a world-constituting field of meaningful experience and action, in which personalities and social systems first attain their identity as individually structured meaning complexes of selected experience and action.[63] With this kind of system identification, it is action that becomes primary (for the reasons already given, i.e., because identification must proceed by way of the attribution of reductions), and a system can be identified on the basis of experience only in so far as it acts. Therefore, the difference between psychic and social systems organizes the ways in which experience and action are attributed and this constitutes what can be observed as psychic and as social system.

In all further considerations here we are treading upon very unstable ground theoretically, so that all I can do is indicate where some of the problems lie. In the case of psychic systems, we have the advantage of important identification aids, viz. the visible unity of the organism to which the psychic system is ascribed and the continuity of the directly experienced life of consciousness with which it knows itself one. Although the meaning system personality first constitutes an I-myself as a unity, this initially knows itself as nonarbitrary and first learns of its contingency from the world, in particular from an insight into the objective-temporal limits of life. Such well-supported and secure identity allows for a close fusion of experience and action, which treats both the action of my own ego and that of another ego as a choosing of fields of experience. The actual carrying out of life, which is motivated in my self and understood in the other, thus has the meaning, prior to all conceptualizing rationality, of a selection of experience—and this in such a way that it is not the chosen experience (which is available to others as well) but only the selection itself that appears as an act and is attributed to the system. It is the act of looking at my hands instead of into my eyes—that I attribute to you. And not what you see, my hand or my eye.

The case with social systems is quite different. They do not have the benefit of such identification aids. They receive their identity only through an intelligible combination of the intended meanings of actions—through the insight that the meanings of action A and action B are somehow connected. It is only as a complex of interrelated actions that they stand out from their environment. Such configurations appear to serve primarily the achievement of reductions that go beyond what the individual consciousness can grasp, i.e., they serve to increase its selection potential. Selective behavior ordered in social action systems is, to the extent that it is ordered there, attributed to the social system and not the psychic system.[64] To this extent, the latter is relieved of responsibility, even though the actions always have their psychic motivation as well. The meaning of such actions—even though they are psychically intended—is borne by the social context and involves no reference (or only a very indirect one) to an individualized psychic system: not-looking-into-my eyes may be prescribed behavior for a servant.

What has been said here does not exclude the possibility of social systems regulating experience. As systems of action, they are themselves experienceable; they require experiencing within the context of the action they order; they eliminate, as reductions, nonpreferred possibilities of experience, or at least make access to distant possibilities more difficult; they may even—e.g., as systems of religious practices or of scientific research—have their primary function in the production of certain contents of experience. For social systems the relationship between experience and action—i.e., as a reminder, between foreign and nonforeign reduction—is extremely complicated and involves both manifest and latent relationships; it is no longer directly "intelligible" and can be uncovered only through multilayered analyses, perhaps those of a "sociology of knowledge." In any case, however, it is fundamentally determined by the fact that a social system attains its identity through meaning relationships among actions and orders experience only per implication of the meaning of such actions.

I shall have to leave the many other branches of this topic unexplored here and return to my starting point, to the question of the use of the concept of meaning in sociological theory formation and of its consequences. This question is obviously of central importance, and decisions taken here will be far-reaching. To the extent that the concept of meaning and the terminology used in explicating it can be made transparent, it will be possible to anticipate what effect such clarifications should have on discussions of both theory and methods in sociology. I want to close here with a few examples.

1. The question of the meaning of meaning—if it is posed as a question of the function and not the existence or essence of meaning—makes possible a radical problematization of the life-world. Problematization here does not mean bringing absolutely everything into doubt—that would be the classical form of a scepticism still laboring under onotological premises—but, instead, finding for every thing that exists a reference problem, on the basis of which it can be looked at for other possibilities. It is only in the world, on a basis of specific conditions of possibility, within the framework of system structures, that other possibilities are possible. Herein lie real and practical limits to meaningful problematization, and ignoring them can lead only to problematization becoming a problem for itself, i.e., becoming reflection.[65] The question of the function of meaning and of the conditions making possibility possible has its radicality not in ignoring such limits, but in the form in which it allows that they may be dealt with; it regards them as *positive*, i.e., as nonnegated constants that constitute a field of problems and possibilities, and thus uses the negation technique already described, which always remains conscious of the negatability of both the negated and the nonnegated.

A sociology that poses the question of meaning in this fashion implies its own dedogmatization. It frees itself from unchanging and uncontrollable ties to a pregiven nature—be this of its object or of its own Reason and its particular conditions for knowledge—and forces itself to constantly make theoretical decisions about which structures it will refrain from problematizing for the sake of which goals of knowledge, i.e., forces itself to take responsibility for itself. Only thus can it fully recognize the fact that its truths are, and remain, hypothetical in nature and that their positivity is nothing other than the structural variability of the system within which it seeks to recognize truths. It must then try, with specifically sociological means, to discover which societal system makes science possible and which science system, sociology; further, what truth means sociologically and which stage in the development of society first makes the interpretation of truth as undeniable intersubjective certainty possible; which social structures are implied in this context by the word *subject* and how society and the world must be ordered if such an abstract interpersonal relationship as that of intersubjectivity is to be practicable; what specific functional contribution can be made in this context by perception, theory, and methods; and not least, which structural conditions make structural variability possible and with it, among other things, the positivity of the conditions for knowledge; and which stage in society's development sets which limits here. That such a sociology of sociology ultimately argues in a circular fashion is obvious. But that need not concern someone who no longer regards science as a foundational exercise in the sense of a logically exact derivation of propositions from invariant first principles, and who instead sees science's achievement in its contribution to the societal constitution of a meaningfully ordered world.[66]

2. Meaning analyses must, then, play a larger explanatory role than the prevailing notion of science would allow them, and we will have to ask how this is methodologically possible. One may assume physical or organic restrictions on meaning formation and locate these in the object or in the subject, but this still does not afford an explanation of particular given contents of meaning. The insight that the contingency of all meaning is an essential functional element excludes any kind of "natural" reductionism, any recourse to noncontingent Being, to final causes, or to an assumed substratum of measurable variables or probabilities. The intersubjectivity of knowledge can no longer be tied to some Given that every reasonable person is capable of discovering. It cannot be secured through empiricism (ultimately, through perception)—that would amount to granting physical things and events a monopoly in mediating among human beings.[67] Instead, what has been referred to here as the intersubjective transferability of ideas and knowledge can be secured only through the form of meaningful experience processing. The question is: how?

At least one answer to this question has been around for a long time. It goes back to the widespread Enlightenment notion that man can have insight only into what he is capable of producing. We gain some insight into this thesis itself if we note that insight had by then already been reduced to necessarily intersubjectively transferable instances of knowledge. What was meant was that a propositional form had to be sought in which the knowing subject posits himself as producer, or more precisely: as initiator of a production process that others are also capable of initiating according to specifiable rules. Interest in this structure, and, by the way, the "technical" aspect of this structure,[68] are not concerned with the mechanics of the production process as such, but rather with the abstraction of the position of subject. Both cognition and production are hereby regarded as independent of all-too-concretely fixed personal attributes—independent of social position and origin, past performance, etc., and thus of the concretely given societal structure as well. Both the pragmatic meaning criterion[69] appearing at the end of the nineteenth century and Max Weber's attempt to rationalize scientific concepts using ideal–typically insinuated means/ends relationships[70] are only later variations of this basic idea.

Today, however, the limits of this technical-pragmatic transferability guarantee are clearly visible. They were first discussed primarily within the historical sciences and hermeneutics, but have more recently come to be seen in attempts to establish a general theory of highly complex systems and in organization theory oriented to decision theory—i.e., not only in the work of Dilthey and Habermas but also that of Bertalanffy, Ashby, or Simon. They are tied to the limits of the classical notion of causality and to the limits of the available logical calculi, which can promise no unequivocal results when applied to highly complex systems and are thus also unable to fully neutralize the subject as decision maker or interpreter. And just such meaning-constituting, contingency-preserving, self-reducing action systems typically fall within this area where the acting subject cannot be abstracted into initiator functions. The question thus arises, especially for sociology, of whether it is not both possible and necessary to broaden this classical form of securing intersubjectivity.

We should consider whether an opportunity to do just this is not given with the concept of function and its accompanying comparative analyses. Functions are problem-related rules of comparability.[71] They hold out the promise of increased knowledge in the form (and *only* in the form) of a comparison of the dissimilar: A and B are functionally equivalent to the extent that they are both able to offer a solution to the problem X, i.e., to transform it into a form for the production of A or of B, each with its own resultant problems. This form bestowal implies a bridging over of material or objective differences—and with it, a bridging over of social differences in situations of experience and action. The differences are thus denied neither in their material, nor in their social dimensions, but are simply overcome by means of a Kantian "insofar" (Sofern) abstraction—the already familiar technique of conditional, conserving *(aufhebende)* negation which can in other instances be negated itself. This makes it possible to also regard the subjects of knowledge as differing yet still interchangeable—to the extent that they understand each other in their selections.

In a functionally conceived system theory, human action does not appear as a causal factor acting or acted upon in accordance with certain laws, but instead as selection that is oriented on meaning, and becomes predictable only by way of system-structural restriction of the possibilities chosen from. The scientific reconstruction of such selections interprets them as choices from among comparable, functionally equivalent problem solutions within the framework of that meaning system on whose structure the problems to be solved depend. The explanation and prediction of action is not exclusively a matter for some science confronting an unknown but nonetheless already determined and, in principle, explainable and predictable reality; it is first and foremost a matter for real action systems themselves and, within these, is something to be balanced against other interests (for example those of adaptation and innovation, of curiosity, excitement, danger). What science accomplishes is only a reconstruction of the rationality of more or less predictable selections. It can in this way offer to action further reaching domains of possibility and so increase its selectivity. But it does not have the function of predicting action, and must instead also be able to reflect the fact that in real action systems the predictability of action is not a goal to be maximized alone.

3. Once we start looking at the implications of meaningful or meaning-based experience processing and expressing them in discussable concepts, it soon becomes obvious that the style of dichotomized controversies which has until now animated the history of sociology and driven it forward is based on assumptions that are far too simple and can no longer be maintained. Such opposing pairs as cooperation and conflict, stasis and change, or norm and fact cannot represent theory alternatives between which sociology must choose. Such situations of choice do exist at the level of individual actions, including those of the theoretician: living and acting require decisions about whether to be prepared in certain instances for cooperation or for conflict; whether to allow for (or work toward) constancy or variability; whether one's expectations will reflect a willingness to learn (cognitive expectation) or an unwillingness to learn (normative expectation). But such decisions

cannot be taken for individual social systems in their entirety and certainly can not be made by the theory of all such systems.

Any reasonably well-equipped sociologist is aware of this anyway, but a clarification of the concept of meaning should make it possible to better see just why this is so. Meaning orders the indication of alternate possibilities, which, while they may be negated, are not totally suppressed or made to vanish completely. All cooperation contains some indication of the possibility of conflict, and this functions as a secret regulative of the forms and conditions of cooperation.[72] For its part, conflict is possible only on the basis of a shared (and even: consciously shared) definition of the situation, something about which there is no conflict. The same is true for the relationship between structurally fixed conditions, on the one hand, and change, on the other. Every meaningful elaboration of structures—at least in the present state of societal development—immediately raises the question of alternate possibilities. Only latency can effortlessly protect structures from reforms or revolutions. Meaningfully established structures, on the other hand, must be accepted and affirmed (meaning: not negated) in the changing flow of experience and action. The difference between facts and norms must also be seen in this way, even though this will seem even stranger to today's sociologists and nonsociologists alike. Meaningfully ordered behavioral expectations may involve prior indication of the possibility of disappointment and, with this, a more or less well-formed idea of what one can do in such a case: persevere or adapt. The expectation, which always remains "factual," takes on a normative or a cognitive style according to which decision has been made and, if expressed, can itself be expected to be expressed in this style.[73] Here as well, sociological theory must refrain from building into its premises what as a meaning-constituting social accomplishment must first be explained. And it will be able to accept this loss to the extent that it comes to understand the special character of meaning-constituted systems and to reflect this in its research plans.

4. One of the difficulties this theory program has to contend with is the old analogy between social system and organism, something that makes it much more difficult to bring out the very special nature of meaning-processing systems. The comparison between social systems and organisms is a product of the European intellectual tradition and, although quite controversial since the nineteenth century, has been a most important impetus for the development of a theory of social systems.[74] The problem here is usually

seen as one of justifying the analogy and establishing its proper limits.[75] The purely metaphorical use of this analogy has since been relinquished. The dominant tendency today is to search for a general theory of all systems, including machines, and to use it in making such a comparison.[76] If this is to succeed, however, it will be necessary to develop the theory of meaning-constituting, i.e., psychic and social, systems to a level comparable to that of the theory of machines or of organisms. My analysis of the concept of meaning should give us an idea of the difficulties involved here.

Previous discussions of the organism analogy have focused, very roughly speaking, on three main differences. The European tradition saw the difference primarily in organisms being made up of interconnected parts, and saw social systems, on the other hand, of parts living separately. More recent criticism of the organism analogy can be expressed as follows: social systems possess a much higher structural variability and are, therefore, not limited to direct exchange processes in securing the requisites of survival from the environment; they can modify these requirements through structural change and are thus able to adapt to a much wider range of possibilities. In addition, there is a third point that has received less attention in this discussion but is emerging very clearly in the general development of sociological theory, viz. that organisms are integrated on the basis of life; social systems, on the basis of meaning. An organism is a living whole which consists of living parts; it cannot be claimed of a social system, however, that it lives as a whole or that it consists of living parts, e.g., persons.[77] We can do justice to the constructive potentialities of social systems only if we think of them as being integrated in a much more abstract fashion, i.e., if we refer them back to bases that are capable of abstraction: they consist not of concrete persons but of meaningfully identified acts.

What we would have to come to understand is this increased capacity for the construction of complex and contingent possibilities and for selective orientation on them. The comparison basis offered by a general systems theory lies in the notion of a system that differentiates structure and process and thereby organize selections. However, organisms, machines, and meaning-constituting systems—i.e., personalities and social systems—all differ in the way they accomplish this, with one fundamental difference being that meaning-based/conscious experience processing makes available a greater—in principle, unlimited—complexity as a selection domain.[78] In a comparison of systems, the special nature of meaning-constituting systems can be brought out by the question of how such a better-organized selectivity is possible.

5. In his work on the theory of the evolution of action systems, Talcott Parsons posed a comparable question about the development of increased and more generalized capabilities of system adaptation.[79] For him, the scheme for answering this question was already indicated by a general theory of action systems and, within that framework, by an interpretation of the relationship between structure and process that increasingly took on the features of a language analogy. With this, the institutionalization of symbolic-cultural "codes" of meaningful behavior became central to evolution theory. In its more advanced stages, that presuppose human beings, evolution is guided by meaning and proceeds—albeit with neither continuity nor absolute necessity—in the direction of generating possibilities of combining increasingly unrestricted ways of adapting to complex and changing environmental conditions. A more exact analysis of the function of meaning-based experience processing and the mechanisms through which it operates would not totally reject this interpretation, but would have to expand it and make it much more complicated, since meaning cannot be adequately grasped solely on the basis of a structure/process difference.

Important contributions can be made here by a general theory of system evolution, a theory whose basic features are already becoming visible. It comprehends evolution as structural change that proceeds in the direction of increased complexity, both on the level of the world as a whole and in some (not all!) systems. This world/system difference is required as a "motor" for evolution: structural changes in individual systems make the environment of other systems more complex, and these react by exhausting new possibilities, or by adaptation or indifference—in any case, by increasing the selectivity of their state. Structural changes beneficial to the adapting system can, in turn, leave the environment of other systems richer in possibilities so that, although complexity does not necessarily increase for all systems or types of systems, it does for their relationship, which is then available to meaningful experience as the world.

In the course of such selective changing of possibilities into actualities, three types of (system-specific) mechanisms are required: mechanisms for the projection of possibilities, mechanisms for the selection of suitable possibilities, and mechanisms for stabilizing what has been chosen in systems. All mechanisms are dependent on system structures, i.e., are developed only within evolution itself. (The theory thus offers no explanation of the beginnings of evolution and only a very general framework for the construction of theories about individual historical sequences.) In the area of organic evolution, new possibilities arise through mutation: a selective component is given by, for example, "the struggle for survival"; and a stabilizing one, granting independence from the fortuity of possibility production, is given by the reproductive isolation of populations. In the case of the organic–psychic system complex man, we might consider as corresponding mechanisms: perception, pleasure/pain differentiation, and memory.[80] For meaning-constituting systems and in particular social systems, no corresponding ideas have been worked out.[81] We still do not have the necessary prior clarification of the special character of such systems, in particular regarding the specific function, and mode of functioning, of meaningful or meaning-based experience processing.

If we see the function of meaning in the preservation of reducible complexity, connecting lines can very easily be drawn to the general theory of evolution. As complexity increases in the course of evolution, everything determinate takes on an increased selectivity —becomes, willingly or unwillingly, a choice from out of an increased number of alternate possibilities. Every yes implies more nos, so that it ultimately becomes advantageous to thematize this implication and make it available in consciousness—in other words, to learn negation.[82] Herein lies the evolutionary significance of the emergence of meaning as a highly involved strategy of processing input from the environment, a strategy that also creates a new kind of system with its own potential for evolution. Since possibilities can then be thematized and negated within meaning itself, meaning makes possible a tremendous increase in the power of all three evolutionary mechanisms and thus, in comparison to physical and organic evolution, makes possible a substantial acceleration of the evolution process. In the meaningfully identified premises of experience processing, it is not only what, in any instance, is actually chosen that is firmly stabilized—i.e., made continually available— but also the whole set of possibilities chosen from. In this way it becomes possible to meet the increasing demand for selection given by an increasingly complex world.[83]

In meaning systems, this advance allows for the onset of new and more rapid processes of evolution which further develop the special potential of meaningfully ordered experience. Possibility production, selection, and stabilization can all be found here in other forms. This can be illustrated by looking at the legal sphere of society.[84] Its development is generally borne by (1) the functional differentiation of society, giving rise in individual functional areas to increasingly daring and increasingly differing norm projections, and thus to a surplus of possibilities for legal regulation; (2) by correspondingly formulated, at first judicial and then legislative, procedures for establishing decisions that select those laws that are to be in force; and (3) by legal dogmas and findings in the form of legal maxims, which stabilize the chosen laws and allow them to be passed on, and which in this way also help control (without strictly determining) which possibilities will be projected and selected. Evolution requires, as these examples show, that these mechanisms remain separate but nonetheless coordinated in the way they work.

I can insert into this general framework those dimension-specific hypotheses about development that I have already referred to briefly. In the material or objective dimension, system structures can be developed from more concrete to more abstract premises of experience processing, a structural change that allows the system involved to accept a more complex environment. We thus typically find, with the transition from archaic societies to the politically integrated civilizations of the ancient world, an abstraction of the religious and moral principles regarded as binding, with the resulting possibility of introducing more varied forms of human behavior into the society and of accepting more individually formed psychic systems. In the temporal dimension, expanding the horizon of possibilities in this way necessitates changes that, in the modern era, involve a mobilization of what can be seen as present, a more complex thematizing and planning of the future, and a "capitalizing" of the past. Meaning-constituting social relationships are, for their part, abstracted in the course of this development and replaced by the formula of the free and equal subject, the constituting consciousness, whose function no longer depends on particular attributes or membership, and whose role may be assumed by anyone. All of these changes allow the meaning of the world to be regarded as contingent and all meaning in the world to be rationalized in terms of its selective function.

That much can be presented as a kind of rough overview or at least as a claim put forward for further discussion. However, my initial thesis of the primacy of negation in meaning-constituting experience will lead us still further. If this thesis is correct, it should be possible to regard evolution at the level of meaning systems as an evolution of the technique of generalizing and reflexive negating. The social system guiding evolution—society—will be able to release the potential for further development only if, in a world that has become extremely complex, it is able to specify alternate possibilities with adequate precision; if it does not absorb discontent in religion or magic or drown it in consumption, but instead is able to redirect it toward attainable alternatives; if it can adequately institutionalize risks; if it can offer enough present security for irregular fluctuations; if it can reverse decisions; and if it can learn—in short: if it can practice a sufficiently differentiated negation.[85] We have little to offer here—certainly no ideas that have been tested in research. It may be that we are still being misled by language or are still in the grips of traditional logic with its all too simple concept of negation. One notices with sociologists in particular just how undifferentiated their negation can be. Perhaps they have not yet devoted enough thought to the concept and the function of meaning.

Endnotes

1. For typical formulations see Paul Hofmann, *Das Verstehen von Sinn und seine Allgemeingültigkeit: Untersuchungen über die Grundlagen des apriorischen Erkennens* (Berlin: 1929). See also Hofmann, *Metaphysik oder verstehende Sinnwissenschaft. Gedanken zur Neugründung der Philosophie im Hinblick auf Heideggers "Sein und Zeit"* (Berlin: 1929). Tendencies to reject the subjective reference of the concept of meaning are unmistakable today. But they end up in a reontification of meaning, a re-fusing of the concepts of meaning and being. This occurs, for example, in the contributions of Hermann Krings and Hedwig Conrad-Martius to Helmut Kuhn and Franz Wiedemann eds., *Das Problem der Ordnung. Sechster Deutscher Kongreß für Philosophie* (Munich: 1960), pp. 125–141 and 141–155, that still encountered considerable opposition; and Max Müller, "Über Sinn und Sinngefährdung des menschlischen Daseins—Maximen and Reflexionen," *Philosophisches Jahrbuch* (1966), 74:1–29.

2. See for this Niklas Luhmann, "Reflexive Mechanismen," *Soziale Welt* (1966), 17:1–23; reprinted in *Soziologische Aufklärung* (Cologne-Opladen: 1970), pp. 92ff.

3. The reason for this is precisely because the history of the word is one of confusions. Parsons uses "constitution" without defining it—see Talcott Parsons, Robert F. Bales, and Edward A. Shils, *Working Papers in the Theory of Action*. New York-London: 1953, p. 44. Husserl uses "constitution" as a central category that remains ambivalent in at least two regards and fluctuates in its meaning, namely between the having of immediate evidence and performance on one hand and between receptive clarification and creative production on the other.

4. In the more abstract classical version (that omits the concept of experience), contingency indicates a being that is capable of not being. Underlying

this is an idea (of a possible being) that can be negated in two ways—with respect to possibility (which then means impossibility) and in the being to which the possibility refers (and this means contingency). Only does this formally determined category acquire the additional connotation of dependence and accidentality—namely with respect to the mode of selection that brings the being that is capable of not being into existence. For the history of the concept see Hans Blumenberg, "Kontingenz," in *Die Religion in Geschichte und Gegenwart* (3d ed.; Tübingen: 1959) vol. 3, col. 1793f. with additional references.

5. See for this James Olds, *The Growth and Structure of Motives: Psychological Studies in the Theory of Action* (Glencoe, Ill.: Free Press, 1956), especially pp. 185ff., for provocative comments on the interconnection between contingency and motivational structure.

6. Cf. for this the remarkable attempt at a *semantic* theory of information by Donald MacKay, "The Place of 'Meaning' in the Theory of Information," in Colin Cherry, ed., *Information Theory: Third London Symposium* (London: 1956), pp. 215–224; as well as MacKay, "The Informational Analysis of Questions and Commands," in Colin Cherry, ed., *Information Theory: Fourth London Symposium* (London: 1961), pp. 469–476. MacKay, too, emphasizes the selective function of meaning but relates it not only to conscious states (experiences) but also to the totality of the conditionally built-in tendencies of an organism(!) to react to the environment. But in this way the, for us, decisive aspect of world preservation remains unclear or assumed as a system-performance in the form of memory etc. See also the critique of such tendencies towards a biological psychology in Talcott Parsons, "The Position of Identity in the General Theory of Action," in Chad Gordon and K. J. Gergen, eds., *The Self in Social Interaction*, vol. 1 (New York: 1968), pp. 11–23.

7. Similarly, we also leave open here the question of the significance that the following remarks possess for logic. It is obvious that they question the treatment of positive and negative judgments as two qualitatively equal basic forms and, with this, the central premise of traditional logic. But a logic that wants to include "ought-judgments" or one that wants to take account of a plurality of originarily experiencing subjects must still pursue such doubt. Cf. also the critical remarks of Edmund Husserl, *Erfahrung und Urteil* (Hamburg: 1948), pp. 352ff., that certainly give primacy to position, not negation, and do not illuminate the genealogy and functionality of negation sufficiently.

8. A noteworthy preparatory work is found in Sigmund Freud's study "Die Verneinung," *Ges. Werke*, vol. 14 (London: 1948; new edition, 1955), pp. 11–15. Freud already viewed negating as a way of rejecting something in itself warranted or proper. But he explained this idea only with respect to problems of the psychical and the organic system, namely with respect to the sublimation and the preservation of an inner/outer difference exemplified in the organic system.

9. Cf. the noteworthy contribution of Ernst Tugendhat, "Die sprachanalytische Kritik der Ontologie," in Hans-Georg Gadamer, ed., *Das Problem der Sprache: Achter Deutscher Kongreß für Philosophie, Heidelberg 1966* (Munich: 1967), pp. 483–493. Of course, we must leave open the question whether linguistic analysis suffices to clarify the genealogy and functionality of negating or whether language does not always presuppose the possibility of negation and merely supplies the unifying symbol and the direct intendability for this very

complex performance. In the final analysis, Tugendhat, too, reverts to an unclarified concept of praxis.

10. Cf. for this distinction Niklas Luhmann, *A Sociological Theory of Law* (London: Routledge & Kegan Paul, 1985), pp. 31ff.

11. It is no accident that the matter is the same in Parsons. He formulates his best statements about meaning in connection with the problem of "internalization." His important insight of a connection between generalization (=indifference to differences of impressions) and the selective organization of an objective connection of aspects of reality is based on the idea of an otherwise impossible "internalization" of the premises of experiential processing. Here and in what follows we replace "internalization" with "constitution" in order to preclude the mistaken idea of importing something from outside. Cf. for all this Talcott Parsons, "The Theory of Symbolism in Relation to Action," in Talcott Parsons, Robert F. Bales, and Edward A. Shils, *Working Papers in the Theory of Action* (Glencoe, Ill.: 1953), pp. 31–62; Talcott Parsons/Robert F. Bales, *Family, Socialization, and Interaction Process* (Glencoe, Ill.: 1955), especially pp. 56f.

12. This distinction also has advantages in an entirely different direction; it makes possible an articulation of the relation of meaningfully conscious experience to the neurophysiological bases of this experience in the organic system, that are similarly, even if in a different way, organized as selective processes. Cf. for this R. M. Bergström, "Über die Struktur einer Wahrnehmungssituation und über ihr physiologisches Gegenstück," *Annales Academiae Scientiarum Fennica*, Series A.V. Medica (1962), 94:1–23; Bergström "Neural Macrostates," *Synthese* (1967), 17:425–443.

13. "Consciousness, and even more so self-consciousness, are information gatherers. And the most encompassing method of gathering information is the one that we call hermeneutic understanding," says Günther in "Kritische Bemerkungen zur gegenwärtigen Wissenschaftstheorie. Aus Anlaß von Jürgen Habermas: "Zur Logik der Sozialwissenschaften," *Soziale Welt* (1968), 19:328–341. In addition to this, Günther, of course, retains the basic and therefore unclearly used concept of reflection. Cf. also Gotthard Günther, *Das Bewußtsein der Maschinen. Eine Metaphysik der Kybernetik* (2d ed.; Krefeld/Baden-Baden: 1963).

14. For a comparison see the distinction between information and meaning in Donald MacKay, "The Place of 'Meaning' in the Theory of Information," in Colin Cherry, ed., *Information Theory: Third London Symposium* (London: 1956), pp. 215–224; and MacKay: "The Informational Analysis of Questions and Commands," in Colin Cherry, ed., *Information Theory. Fourth London Symposium* (London: 1961), pp. 469–476. MacKay also believes that the need for a selective performance—but related to the states of the organism—is the beginning of the distinction. Of course, he does not go beyond a distinction between the selective function (meaning) and its actual performance (information).

15. An example may help to clarify this: at the dinner table it becomes clear that the children have not washed their hands. This may be for the father, whose learning is attuned cognitively, an experience-confirming, anticipatable normality without essential informational value; for the mother, who anticipates things in a morally normative fashion, deviant behavior that is always new and informative; for the children themselves who actually live through this experience a completely unexpected surprise with great relevance and corresponding informational content.

16. Cf. for this Hans-Georg Gadamer, *Wahrheit und Methode: Grundzüge einer philosophischen Hermeneutik* (Tübingen: 1960), pp. 335ff. English edition: *Truth and Method*, G. Barden and J. Cumming, trans. (London: 1975).

17. This statement refers—although it cannot be discussed here in more detail—to a concept of science understood as a social system that is differentiated for the purpose of the learning of cognitive expectations and whose "positivity" resides in the fact that it can treat cognitive structures as variable and can therefore exploit experiences structurally and not merely normatively.

18. For statements of this type see, for example, Alfred Kuhn, *The Study of Society: A Unified Approach* (Homewood, Ill.: 1963), p. 152; or Karl W. Deutsch, *The Nerves of Government: Models of Political Communication and Control* (New York-London: 1963), p. 77. See also my own unclear formulations in Niklas Luhmann, *Funktion und Folgen formaler Organisation* (Berlin: 1964), pp. 191f.

19. Similar interpretations, based on Mead and Dewey, are found in Tamotsu Shibutani, *Society and Personality: An Interactionist Approach to Social Psychology* (Englewood Cliffs, N.J.: 1961), pp. 139ff. (especially p. 148); and Shibutani, *Improvised News: A Sociological Study of Rumor* (Indianapolis: 1966).

20. Especially in Jürgen Frese, "Sprechen als Metapher für Handeln," in Hans-Georg Gadamer, ed., *Das Problem der Sprache. Achter Deutscher Kongreß für Philosophie, Heidelberg 1966* (Munich: 1967), pp. 45–55. Cf. also Siegfried J. Schmidt: "Zur Grammatik sprachlichen und nichtsprachlichen Handelns. Sprachphilosophische Bemerkungen zur soziologischen Handlungstheorie von Jürgen Habermas," *Soziale Welt* (1968), 19:360–372. One readily concedes the attractiveness of this idea: to understand action from the point of view of its most complex, not its simplest, form.

21. Recent investigations of the processing of surprising phenomena in daily life are helpful in this case. See, for example, Charlotte G. Schwartz, "Perspectives on Deviance. Wives' Definitions of their Husbands' Mental Illness," *Psychiatry* (1957), 20:275–291; Fred Davis, "Deviance Disavowal: The Management of Strained Interaction by the Visibly Handicapped," *Social Problems* (1961), 9:120–132; Harold Garfinkel: "Studies of the Routine Grounds of Everyday Activities," *Social Problems* (1964), 11:225–250; Marvin B. Scott/Stanford M. Lyman, "Accounts," *American Sociological Review* (1968), 33:46–62.

22. Cf. For this the distinction between "complexity in time" and "complexity in form" that is important for the constitution of organisms as well as for the programming of machines in J. W. S. Pringle, "On the Parallel Between Learning and Evolution," *Behaviour* (1951), 3:174–215 (pp. 184f.), reprinted in *General Systems* (1956), 1:90–110.

23. In semantic research this interpretation is approximated less by the pragmatic-behavioristic accounts of the Americans than by the Marxist-oriented efforts of Adam Schaff, *Einführung in die Semantik*, German translation (Berlin: 1966). Schaff moves away from a one-sided hypostatization of the (objective) symbolization *[Zeichenhaftigkeit]* of meaning and tries to consider the communicative (social, societal) as well as the historical (temporal, genetic) aspects of meanings. In this way he assumes that the integration of these different dimensions of Marxist theory has already been achieved.

24. A suspicion that made the epistemologically thinking Max Adler speak of the "social apriori of society." See especially *Das Rätsel der Gesellschaft: Zur erkenntnis-kritischen Grundlegung der Sozialwissenschaften* (Vienna: 1936).

25. Cf. Kurt Goldstein and Martin Scheerer, "Abstract and Concrete Behav-

ior: An Experimental Study with Special Tests," *Psychological Monographs* (1941), vol. 53, no. 2, partially translated under the title "Die Unterscheidung konkreten und abstrakten Verhaltens," in Carl F. Graumann, ed., *Denken* (Cologne-Berlin: 1965), pp. 147–153; also O. J. Harvey/David E. Hunt/Harold M. Schroder, *Conceptual Systems and Personality Organization* (New York-London-Sydney: 1961); Harold M. Schroder, Michael J. Driver, and Siegfried Streufert, *Human Information Processing: Individuals and Groups Functioning in Complex Social Situations* (New York-London-Sydney: 1967); Robert Ware and O. J. Harvey, "A Cognitive Determinant of Impression Formation," *Journal of Personality and Social Psychology* (1967), 5:38–44.

26. We will leave open here the much discussed question of the possibility of knowing another ego as another ego and its experience as experience without having direct access to the consciousness of their experiences. Apparently, this interpretation is so advantageous for explaining otherwise irregular and unforeseeable impressions about the behavior of others that it is irresistible and is accepted with a high degree of probability. Its "verification" is a result of the successful interpretations that are attainable through it, not of the (impossible) participation in the conscious life of another.

27. Among the many analyses of this process see particularly George H. Mead, *Mind, Self and Society From the Standpoint of a Social Behaviorist* (Chicago: 1934), which includes the term "role-taking" that had a very misleading effect on the subsequent use of the concept of "role." See also Mead, *The Philosophy of the Act* (Chicago: 1938), and "The Objective Reality of Perspectives," *Proceedings of the Sixth International Congress of Philosophy 1926* (New York: 1927), pp. 75–85. See also Alfred Schütz, *Der sinnhafte Aufbau der sozialen Welt: Eine Einleitung in die verstehende Soziologie* (Vienna: 1932), with a concept of meaning that suggests that reflection is still "essentially" bound to the subject.

28. This results in a fundamental objection against the hypostatization of a unifiedly transcendental subject. Such a subject could not, in principle, experience objectively because it would lack a horizon of ready perspectives that, at any moment, are not its. One would have to attribute objectivity to it only as an all-encompassing, nonselective experience and this would destroy every possible interpretation of the concept of experience. It seems to be a consequence of this interpretation that the transcendental subject tends to lose its subjectivity and to dry up into a connection of rules applied by a plurality of subjects.

29. Cf. Jean Cazeneuve, "La connaissance d'autrui dans les sociétés archaïques," *Cahiers internationaux de sociologie* (1958), 25:75–99.

30. See Peter Berger and Hansfried Kellner, "Die Ehe und die Konstruktion der Wirklichkeit: Eine Abhandlung zur Mikrosoziologie des Wissens," *Soziale Welt* (1965), 16:220–235.

31. See Niklas Luhmann, "Selbststeuerung der Wissenschaft," *Jahrbuch für Sozialwissenschaft* (1968), 19:147–170 (149ff.); reprinted in *Soziologische Aufklärung*, pp. 232ff.

32. See for this also Erich Feldmann, "Versuch einer Theorie der Gegenwart," *Festschrift für Erich Rothacker* (Bonn: 1858), pp. 131–146; Klaus Held, *Lebendige Gegenwart* (The Hague: 1966).

33. Even the dead, as long as they are not experienced as co-experiencing subjects (as is typical in many early societies), are not left behind in the past—

as it were, looking upon our past as their present—but are taken along in the continuing present because this is the only way in which they can be of current interest.

34. See the corresponding analysis in Schütz, pp. 111.

35. Regarding this neutralization of our concept of time, cf. Helmut Plessner, "Über die Beziehung der Zeit zum Tode," *Eranos Jahrbuch* (1952), 20:349–386.

36. For the sake of clarification it should be mentioned that the customary characterization of archaic experience in terms of its being bound to tradition seems to me to miss the point or merely to grasp a secondary characteristic. More important is the overwhelming preeminence accorded the present in which life must take place and whose existence (that is rich in risks but poor in possibilities) provides the occasion to look for security in the repetition of the past. Innovation is by no means excluded. But it is admitted only if it can be stabilized quickly and successfully in the present. See Siegfried F. Nadel, "Social Control and Self-Regulation," *Social Forces* (1953), 31:265–273; and for the specific area of law, the excellent work of Louis Gernet: "Le temps dans les formes archaïques du droit," *Journal de psychologie normale et pathologique* (1956), 53:379–406.

37. Seen from this point of view one can understand the genuine prominence attained by concepts like *capital* and *history* in the nineteenth century. They permit a distancing from a past that is still powerful in daily life; permit the constitution of meaning in a time-relation that is radically changed by societal evolution.

38. In Marshall D. Sahlins and Elman R. Service, eds., *Evolution and Culture* (Ann Arbor: 1960), p. 38.

39. I owe this idea to Stefal Jensen, *Bildungsplanung als Systemtheorie* (Bielefeld: 1970), p. 67.

40. An interpretation of Weber with this in mind is presented by Horst Baier, Von der Erkenntnistheorie zur Wirklichkeitswissenschaft: Eine Studie über die Begründung der Soziologie bei Max Weber, Habilitationsschrift (Münster: 1969), (unpublished).

41. Basically, this is present already in Ferdinand Tönnies, "Zweck und Mittel im sozialen Leben," *Erinnerungsgabe für Max Weber* (Munich-Leipzig: 1923), 1:235–270.

42. In more recent sociopsychological and sociological research this idea is expressed as an opposition between instrumental, task-related and expressive, emotional, consumatory variables in social systems. To be sure, the opposed terms of expression, feeling and consumption are one-sided and insufficient in different ways. But together they illuminate the culturally suspect, residual character of the immediate presence of experience very well.

43. This idea is applied to the theme of trust in Niklas Luhmann, *Vertrauen: Ein Mechanismus der Reduktion sozialer Komplexität* (Stuttgart: 1968), especially pp. 9ff. English version: T. Burns and G. Poggi, eds.; H. Davis et al. trans., *Trust and Power* (Chichester: 1979).

44. For the time being, detailed ideas about this exist only in psychology. Above all cf. O. J. Harvey/David E. Hunt/Harold M. Schroder, *Conceptual Systems and Personality Organization* (New York-London: 1961).

45. To be exact, even simple contingency is already a complicated state of affairs. My expectations can be disappointed because I do not prepare myself for my experiences (e.g., do not go and see for myself) and because the world

frustrates me. James Olds, p. 185ff., already calls this "double contingency" and views social contingency only as a specific case of it. In this text we will follow the better-known terminology of Parsons. See, for example, Talcott Parsons and Edward A. Shils, eds., *Toward a General Theory of Action* (Cambridge, Mass.: 1951), p. 16.

46. Parsons himself did not clearly work out the consequence of his formulation of this problem and therefore ran the risk of construing the concept of social structure purely behavioristically in terms of expectations of sanctions. This is the reason why, for him, the normativity of social structure remained an insufficiently verified postulate that always provoked critique. Important analyses that go beyond this and explicitly include the level of the expectation of expectations are found in Johan Galtung, "Expectations and Interaction Processes," *Inquiry* (1959), 2:213–234; Thomas J. Scheff, "Toward a Sociological Theory of Consensus," *American Sociological Review* (1967), 32:32–46; and above all, Ronald D. Laing, Herbert Phillipson, and A. Russell Lee, *Interpersonal Perception: A Theory and a Method* (London and New York: 1966).

47. The true function of meaning resides for many—as it does for Jürgen Habermas, *Erkenntnis und Interesse* (Frankfurt 1968)—in deciphering the, in itself, inaccessible experience of others. This is one-sided insofar as meaning, and even the expectation of expectations, can also be used to "by-pass" the subjectivity of others without looking in.

48. Consequently, it belongs to the most general rules of social tact not to introduce unwelcome themes before one can expect the expectations of one's partner.

49. See for the case of conflict, for example, John P. Spiegel, "The Resolution of Role Conflict Within the Family," *Psychiatry* (1957), 20:1–16; or Thomas C. Schelling, *The Strategy of Conflicts* (Cambridge, Mass.: 1960), especially pp. 54ff.

50. More on this can be found in Niklas Luhmann, *A Sociological Theory of Law*.

51. Examples of distinct levels of development are perhaps: transference of the risk of defects to the seller of a commodity in the interest of commerce (Roman law) or today: legal claims to compensation for the destruction of apples in a determinate harvest year.

52. *Über Sinn und Sinngebilde: Verstehen und Erklären* (Tübingen: 1929), pp. 54ff.

53. For example, see Schaff, pp. 175ff.

54. This seems to be suitable as a starting point for discussion of the societal conditions of art. To observe how this can be exploited for a theory of public opinion see Niklas Luhmann, "Öffentliche Meinung," *Politische Vierteljahresschrift* (1970), 11:2–28.

55. Among others cf. Talcott Parsons/Edward A. Shils, (eds.), *Toward A General Theory of Action* (Cambridge, Mass.: 1951), pp. 108f.; Talcott Parsons, *The Social System* (Glencoe, Ill.: 1951), p. 482; P. G. Herbst, "A Simple Theory of Behavior Systems," *Human Relation* (1961), 14:71–94, 193–239 (78ff.); Alfred Kuhn, *The Study of Society: A Unified Approach* (Homewood, Ill.: 1963), especially pp. 48ff.; Gabriel A. Almond, "A Developmental Approach to Political Systems," *World Politics* (1965), 17:183–214 (187ff.); David Easton, *A Framework for Political Analysis*, (Englewood Cliffs, N.J.: 1965), pp. 24f., 60ff.; Daniel Katz and Robert L. Kahn, *The Social Psychology of Organizations* (New York-London-Sydney: 1966), pp. 60ff, 122ff. In a lecture (1960) Parsons had explained that a

"boundary" is the twofold fact that [1] it is pure chance if internal and external facts coincide and that [2] the stability of internal states is independent of changes in the relations between internal and external states.

56. In the following we will omit the necessary related remarks concerning the case of psychical systems of personalities in order to avoid needlessly complicating the presentation.

57. One rarely finds discussions that are thematically directed to this question. Cf. notably Donald T. Campbell, "Common Fate, Similarity, and Other Indices of the Status of Aggregates of Persons as Social Entities," *Behavioral Science* (1958), 3:14–25, which relies on the preliminary work of gestalt psychology.

58. Very characteristic of this are recent formulations that hold that actions *are* systems—e.g. (1968), p. 14: " 'Action' I define as a system of behavior of living organisms which is organized—and hence controlled—in relation to systems of cultural meaning at the symbolic level."

59. And not accidentally with the help of the idea of the owner whose legally protected freedom of disposition makes such an attribution of action possible. See for this, and for such a historical relativization of conceptions of action *as such*, Friedrich Jonas, "Zur Aufgabenstellung der modernen Soziologie," *Archiv für Rechts- und Sozialphilosophie* (1966), 52:349–375.

60. Among other things, this would require the presentation of complexity as a temporal succession of selective steps that is discussed above.

61. For the sake of clarity it should be remarked that this holds only for meaning-constituting systems. And that for organisms or machines the same problem of performance-related system-identification has to be solved by different conceptual means.

62. A system-theoretical analysis of wholes that comes very close to our argument is found in Andras Angyal, "The Structure of Wholes," *Philosophy of Science* (1939), 6:25–37, with a concept of "system" that is certainly too narrow for our purposes—because it is defined by one principle alone.

63. For comparable remarks see, for example, J. Milton Yinger, "Research Implications of a Field View of Personality," *American Journal of Sociology* (1963), 68:580–592; Talcott Parsons, "Levels of Organization and the Mediation of Social Interaction," *Sociological Inquiry* (1964), pp. 207–220. This is not to be confused with a misleading psychological reductionism whose representatives typically underestimate the system character (and with it the complexity) of the object of psychology and therefore believe that psychology can offer comprehensive theories of individual behavior of a higher degree of abstraction than sociology. Typical representative are, e.g., George C. Homans, particularly clearly in "Bringing Men Back," *American Sociological Review* (1964), 29:808–818; Andrzej Malewski, *Verhalten und Interaktion: Die Theorie des Verhaltens und das Problem der sozialwissenschaftlichen Integration* (Tübingen: 1967); Hans Albert, "Erwerbsprinzip und Sozialstruktur: Zur Kritik der neoklassischen Marktsoziologie," *Jahrbuch für Sozialwissenschaft* (1968), 19:1–65.

64. This results in the important hypothesis that, in the case of the same selective performance—thus in the case of the same world complexity—structurally indeterminate, performatively weak social systems require personalities that are more complex; in other words, that in a complex world social systems can be structured only indeterminately if one can assume the corresponding psychical capacities. See Paul Stager, "Conceptual Level as a Composition

Variable in Small-Group Decision Making," *Journal of Personality and Social Psychology* (1967), 5:152–161; and, in general, research that investigates the social conditions of psychical creativity.

65. Such reflection can assume the strict form of reflexive problematization of problematizing and then lead to the above treated question of the availability of boundaries. For the most part, however, it is documented only in the superficial application of euphemistic characterizations of that state, like "subject" or "will" or "freedom" or "revolution" or "democracy."

66. See the similar position of Georg Simmel, *Soziologie. Untersuchungen über die Formen der Vergesellschaftung*, 2d ed. (Munich-Berlin: 1922), pp. 1ff., that is still formulated in the language of transcendental epistemology, however; and in reference to this Karin Schrader-Klebert, "Der Begriff der Gesellschaft als regulative Idee: Zur transzendentalen Begründung der Soziologie bei Georg Simmel," *Soziale Welt* (1968), 19:97–118.

67. In this way the role of perception or experience (=expectationally structured perception) is not negated in the scientific process of knowledge. This would be absurd. But, henceforth, the question of the function of perception can no longer be answered metaphysically as the conferral of being but itself requires further investigation. A way of providing an answer presents itself if one considers that, among all conscious functions, perception possesses the greatest potential for actual complexity and is determined by this to show the way. This is connected with the fact that, unlike in the case of perceptions, one almost never experiences disappointments in operating with concepts so that stimuli to learn come primarily from experience. Both these aspects are hardly considered by the dominant, supposedly empirical, methodology or, in any event, not systematically exploited. The latter's interpretation of perception as a being-related arbiter among many possible conceptual notions would have to be reversed completely in order for experience, in accordance with its specific nature, to retain the function of providing a field of possibilities from which selections then could be made conceptually.

68. A good example of this can be found in Hans Blumenberg, *Lebenswelt und Technisierung unter den Aspekten der Phänomenologie* (Turin: 1963).

69. Cf. for this Jürgen Habermas, *Erkenntnis und Interesse*, (Frankfurt: 1968), pp. 143ff.

70. See Horst Baier, "Von der Erkenntnistheorie zur Wirklichkeitswissenschaft: Eine Studie über die Begründung der Soziologie bei Max Weber" (Habilitationsschrift Münster: 1969; unpublished).

71. See, for this (not universally accepted), interpretation of functional analysis Niklas Luhmann, "Funktion und Kausalität," *Kölner Zeitschrift für Soziologie und Sozialpsychologie* (1962), 14:617–644; reprinted in *Soziologische Aufklärung*, pp. 9ff.

72. For the domain of organized social systems this is clearly revealed in the fact that the possibility of joining the organization and leaving it becomes the principle that determines the constitution of formal structures. As a development of this idea, see Niklas Luhmann, *Funktionen und Folgen formaler Organisation* (Berlin: 1964).

73. Cf. for this Johan Galtung, "Expectations and Interaction Processes," *Inquiry* (1959), 2:213–234; and Niklas Luhmann, "Normen in soziologischer Perspektive," *Soziale Welt* (1969), 1:28–48. A more detailed presentation is found in Niklas Luhmann, *A Sociological Theory of Law*.

78 MEANING AS BASIC CONCEPT

74. From the more recent discussion, see Corrado Gini, *Organismo e società* (Rome: 1960); Paul Kellermann, *Kritik einer Soziologie der Ordnung: Organismus und System bei Comte, Spencer und Parsons* (Freiburg: 1967); A. James Gregor, "Political Science and the Uses of Functional Analysis," *The American Political Science Review* (1968), 62:425–439.

75. This by no means automatically occurs to a person if, for the sake of comparison, they bring in the relation of "organism" and *psychical* system. In this case, namely, the psychical system based on meaning (personality) is not compared with an organism in terms of isomorphy. Instead, it is viewed only as a steering level, as a program of an organism and not as an independent system. Talcott Parsons, in particular, interprets this quite differently in a recently developed distinction between individual personality and behavioral organism as different subsystems of a general system of action. See, for example, Talcott Parsons, "The Position of Identity in the General Theory of Action," in Chad Gordon and Kenneth J. Gergen, eds., *The Self in Social Interaction*, (New York: 1968), 1:11–23.

76. See, for example, Daniel Katz and Robert L. Kahn, *The Social Psychology of Organizations* (New York: 1966), especially pp. 30ff; Walter Buckley, *Sociology and Modern Systems Theory* (Englewood Cliffs, N.J.: 1967); Anatol Rapoport, "Mathematical, Evolutionary, and Psychological Approaches to the Study of Total Societies," in Samuel Z. Klausner, ed., *The Study of Total Societies* (Garden City, N.Y.: 1967), pp. 114–143 (119ff.).

77. This is the point at which modern sociological systems theory differs from the old European, ethicopolitical theory of society. The latter had always begun from man, indeed from man interpreted in a peculiarly natural-ethical way. And it defined the distinctive, inimitable form of humanity precisely in the fact that it understood the relation of social systems to man in terms of whole and parts. From this point of view, social systems were obliged to establish the good life for its parts (namely human beings).

78. This statement does not necessarily imply the other: that social systems *are* more complex than, perhaps, personalities or organisms (as, for example, on the basis of the old assumption that social systems are constituted out of human beings and because of this are higher, more complex organisms); see René Worms, *Organisme et société* (Paris: 1895), pp. 7ff., 75ff., and many others. A comparison of the complexity of such different systems would presuppose as yet uninvestigated possibilities of measurement. Besides, there is much that speaks for the opposite argument: that meaning is an evolutionary achievement that makes greater world-complexity available on the basis of simpler systems.

79. See, in particular, Talcott Parsons, "Evolutionary Universals in Society," *American Sociological Review* (1964), 29:339–357, reprinted in Parsons, *Sociological Theory and Modern Society* (New York: 1967), pp. 490–520; Parsons, *Societies: Evolutionary and Comparative Perspectives* (Englewood Cliffs, N.J.: 1966).

80. In this case, one does not find the corresponding investigations under the title of evolution theory but in learning theory. At present this parallel is consciously followed. See Pringle (1951) and Donald T. Campbell, "Methodological Suggestions from a Comparative Psychology of Knowledge Processes," *Inquiry* (1959), 2:152–182.

81. Occasionally, one finds passing references. See, for example, Alvin Boskoff, "Functional Analysis as a Source of a Theoretical Repertory and Research

Tasks in the Study of Social Change," in George K. Zollschan and Walter Hirsch, eds., *Explorations in Social Change* (London: 1964), pp. 213–243 (224ff.). Boskoff interprets the mechanisms of evolution as "phases" of evolution. The time relation, however, is less important than the fact that this is a matter of reciprocal relations that increase complexity and that mutually presuppose each other.

82. Besides, this is where one can see one of the transcendental trip-wires that snares anyone who tries to find a presuppositionless foundation: when we speak about phases of evolution before the invention of meaning—indeed even try to imagine them—we have to impute meaning to them and at the same time negate this imputation for the systems of that time. We meaningfully imagine our past as a present that contained no meaning and that, for temporal as well as objective reasons, is inaccessible to us. But our mastery of the technique of negating still makes it possible for us to conceive its non-presence in the form of negation.

83. This does not assert the existence of necessary historical laws of evolution; merely that achievements that presuppose more as well as perform more are capable of being stabilized. Of course, in an increasingly complex world there are also other kinds of alternatives—like a stone's preserving its indifference to an increasingly complex environment.

84. See Niklas Luhmann, "Die Evolution des Rechts," *Rechtstheorie* (1970), 1:3–22. See *A Sociological Theory of Law*.

85. It is very similar in Magoroh Maruyama, "Goal-Generating Dissatisfaction, Directive Disequilibrium, and Progress," *Sociologia Internationalis* (1967), 5:169–188, for progress in the cybernetic system depends on the development of relevant, specifiable, innovative dissatisfaction and therefore usually is codetermined by the religious and moral forms of the social control of deviation, anxiety and dissatisfaction.

3.
Complexity and Meaning

It has been the norm in the past to insist on differences between the science and the humanities, or *Naturwissenschaften* and *Geisteswissenschaften*. This is, in part, a reaction to the marvellous successes of the "real sciences." If the humanities cannot show similar results, then, it is assumed that it is because they have a different identity. The famous expression, the "two cultures"—similar to that of the "two nations" of capitalists and workers of the nineteenth century—has become a habit of thinking, backed by collegial respect of the sort one has for something that one does not understand. Of course, we also have a unity of science movement, but this is a reaction to the previous split of the intellectual field and, by the very fact of being only a reaction, the unity remains weak and the split strong. We live upon difference, not upon unity.

The concepts that serve as the title of my essay seem to mark this difference. They indicate the core problem of the two knowledge groups and of their two different types of research. These two knowledge groups, today, have become aware of seemingly insolvable problems, and no longer identify with a specific subject matter or domain of research but rather with a problem: complexity for the sciences, meaning for the humanities. Overstating the point a little, we could say that the hard sciences identify with complexity and the soft sciences with meaning. The core problem of the hard sciences is the complexity of complexity and that of the soft sciences is the meaning of meaning.

But are these really different issues? Or is it simply as a consequence of our habit of separating the "two cultures" that we distinguish between these two issues? Moreover, there seems to be a change in the way these two knowledge groups identify themselves. They no longer have their own subject matter or domain. This kind of atomistic, "last element" orientation has disappeared in science, and if some of the humanities still stick to their own type of objects, they only demonstrate their own backwardness. The most advanced fields of these two knowledge groups identify themselves by the way they solve insolvable problems.

Theories of decision making and planning, computer programming, research, methodology, and cybernetics and systems analysis run into the complexity of complexity; on the other hand, hermeneutics, theology, jurisprudence, pedagogy, and similar disciplines deal with the meaning of meaning. However, if the assumption is true that the identity of a research field consists in its particular way of solving insolvable problems, in its special forms of logical uncleanliness and dishonesty, and in its way of coping with its fundamental paradox, then it becomes so much more important to ask whether the paradoxes of the two knowledge groups are really different paradoxes, why, and in what sense.

When thinking about complexity, two different concepts come to mind. The first is based on the distinction between elements and relations. If we have a system with an increasing number of elements, it becomes increasingly difficult to interrelate every single element with every other element. The number of possible relations becomes too large with respect to the capacity of elements to establish relations. We can find mathematical formulas calculating the number of possible relations, but every operation of the system that establishes a relation has to choose one among many—*complexity enforces selection*. A complex system comes about by selection only. This very necessity of selection *qualifies* elements, that is, it gives quality to pure quantity. Quality in this case is nothing but limited selective capacity; it is negentropy compared to entropy—entropy meaning that all logically possible relations have an equal chance of realization.

This concept of complexity is based on the concept of operation. It is the *complexity of operations*. The other concept is defined as a problem of *observation*. Now, if a system has to select its relations itself, it is difficult to foresee what relations it will select, for even if a particular selection is known, it is not possible to deduce which selections will be made. Knowledge of one element does not lead to knowledge of the whole system; the observation of other elements will, however, give additional information about the system. The complexity of the system, from this perspective, is a measure for lack of information. It is a measure for negative redundancy and for the uncertainty of conclusions to be drawn from actual observations. By operation, I mean the actual processing of the reproduction of the system. In dynamic systems—which consist of their operations—operations and elements become indistinguishable. Moreover, in *autopoietic systems,* everything that is used as a unity by the system, including the elementary operations, is also produced as a unity by the system.

By observation, on the other hand, I mean the act of distinguishing for the creation of information. Often a distinction is made between internal and external observation: this distinction is not needed, however, since the concept of observation includes self-observation. For example, within the communicative system of this meeting, we can say that this lecture is not easy to understand; we normally use "indexical expressions" when engaging in self-observation of the social system.

The relation between operation and self-observation constitutes a major problem. In particular, it has to be decided whether self-observation (or the capacity to handle distinctions and to process information) is a prerequisite of autopoietic systems, and if so, whether every particular operation of such a system requires a corresponding observation for controlling its selectivity. And again, if so, whether this observation has to include a self-observation of the operation itself, with the operation then serving as identification of self-identity and self-diversity (to use a formulation of Alfred North Whitehead). I cannot elaborate on these problems, for I wish to pursue a more limited goal. My point is simply that both notions of complexity, based on operation and observation respectively, focus upon enforced selectivity. Complexity thus means that every operation is a selection, whether intentional or not, whether controlled or not, whether observed or not. Being an element of a system, an operation cannot avoid bypassing other possibilities. Only because this is the case can we observe an operation selecting a particular course to the exclusion of others. And only because operations can be observed, self-observation becomes possible (be it necessary or not as a requirement of the operation itself). Enforced selectivity is the condition of the possibility for both operation and observation. Further, enforced selectivity is the core problem that defines complexity as a problem for both operations and observations. The latter statement is at the basis of my contention that *meaning is nothing but a way to experience and to handle enforced selectivity,* as I will discuss below.

The best way to approach the meaning of meaning might well be the phenomenological method. This is by no means equivalent to taking a subjective or even a psychological stance. On the contrary, phenomenology means: taking the world as it appears without asking ontological or metaphysical questions. American and "phenomenological" sociologists do not usually make this distinction. It is, nevertheless, essential.

The phenomenological approach describes reality as it appears. Whatever appears can be interpreted as being the exclusion of other possibilities. It may not be what it seems to be, but its selectivity cannot be denied. This epistemological view is similar to that of structuralism: if something appears as structure (or can be observed or reconstructed as structure), it is a strong argument for its being an indication of reality.

Now, if this seems to be a promising route to a forever unknown reality, how can we describe the phenomenon of meaning? Meaning always involves focusing attention on one possibility among many; William James spoke of "fringes," Edmund Husserl of "Verweisungen" into the infinite. There is always a core that is given and taken for granted which is surrounded by references to other possibilities that cannot be pursued at the same time. Meaning, then, is actuality surrounded by possibilities. The structure of meaning is the structure of this difference between actuality and potentiality. Meaning is the link between the actual and the possible; it is not one or the other.

But this is only a first approximation. We have to take into account that meaning presupposes dynamic autopoietic systems— either psychic systems using consciousness as the medium of their operations or social systems using communication as the medium of their operations. In both cases the basic elements are not stable units (like cells or atoms or individuals) but events that vanish as soon as they appear. Their continuous reproduction continuously requires new elements. They cannot accumulate elements—actions, for example—because the continuous disappearance of their elements is a necessary condition of their very continuity. Thus, dissolution and disintegration become a necessary cause of their reproduction. Without the continuous loss of all of their elements, these systems would, after a short span of time, preserve too many elements without then being able to preserve order.

Therefore, meaning has to be based on the instability of elements. This is just another way of saying that meaning is a property of dynamic systems only. This basic precondition reappears in what we can call the instability of actuality. The focus of actual meaningful experience cannot stay where it is, it has to move. The structure of meaning based on the difference between actuality and potentiality relates to this problem. The function of its dual structure is to organize alternating attention: actuality that is certain but unstable, and potentiality that is uncertain but stable. Indeed, we have to pay for our world either with instability or with uncertainty; we have no access to stable certainty. However, we can improve the situation by relating the inverse problems of unstable certainty and stable uncertainty. This relation may come out as meaning and it may evolve with the variation and cultural selection of successful meanings. This evolution of meaning seems to result in increasing complexity.

Complexity has been characterized by enforced selectivity. What we call "organized complexity" or structured complexity seems to evolve as an attempt to direct, or at least limit, the selectivity of operations, structure being nothing but the selection of selections. Meaning is a different articulation of the same problem. Meaning can be considered as an evolutionary universal, giving a new and powerful form to the old problem of complexity. Complexity appears as the world—the ultimate horizon (to again use Husserl's terminology)—of other possibilities accessible from every actual experience. However, since actuality shifts from instant to instant, it requires operations to select the next focus of attention. The world is appresented (Husserl) to every concrete item. It remains, however, inaccessible. It remains the horizon of operations that moves as they move. In this sense, the world of meanings represents enforced selectivity and is characterized by indeterminate determinability. And since we cannot transcend meaning, since we cannot leave the meaningful world in a meaningful way, and since every negation of meaning presupposes meaning, we have no choice but to accept and process enforced selectivity.

My conclusion, therefore, can be expressed by saying: *meaning is a representation of complexity*. Meaning is not an image or a model of complexity used by conscious or social systems, but simply *a new and powerful form of coping with complexity under the unavoidable condition of enforced selectivity*.

Now, if this solves our problem of the two cultures, what does it cost? What kind of expectations will be disappointed? If we can reduce the problem of complexity and the meaning of meaning to a common fundamental problem of enforced selectivity, what are the consequences of this view for the so-called soft sciences, or the humanities?

First of all, it needs to be pointed out that there is no way to deduce structures or essential variables from such an approach. We may, however, develop different theories that are consistent with the theorem of enforced selectivity. Even theologians could feel at home with this approach, if they remind themselves of the old problem of the necessity of contingency according to which even God could not avoid committing himself to creating the world.

However, the most important conclusions can be drawn if we decide to desacralize the theory and to reformulate what has been called *creatio continua* as a completely mundane affair, namely as the construction and reproduction of order out of order and disorder. Then, a distinction has to be made between life and meaning as quite different levels of system building. Meaning itself operates on two different levels, using consciousness or communication as media. Living systems are based on life, psychic systems on consciousness, and social systems on communication. Conscious systems are not living systems, and social systems are not conscious systems; each of them requires the other, however, to be part of its environment. Each of them may be a dynamic and even autopoietic system able to combine closure and openness; but since they are based on different elements, they cannot be part of one encompassing autopoietic system. In any case, even if we can conceive of society as an autopoietic social system, consisting of communications and reproducing communications by communications, it will not be a living system like the *kósmos* of Plato, or a conscious system like Hegel's *Geist*.

Bad signs, then, for the humanities in the old sense and also for what in France is called "sciences de l'homme." We will have to choose system references that crosscut the unity of the human being. We may very well continue to observe human beings as entities behaving in an environment, but we may have to acknowledge that this is not the perspective that leads us to understand how meaning operates in a complex world.

4.
The Improbability of Communication

Without communication there can be no human relations, indeed no human life. Communication theory cannot therefore be confined to examining only certain sectors of life in society. It is not enough to engage in exhaustive discussion of particular techniques of communication, even though, because of their very novelty, such techniques and their consequences are attracting special attention in contemporary society. It is equally inadequate to begin with a discussion of concepts.[1] That would serve a useful purpose only if one already knew what the concept was intended to achieve and in what theoretical field it was to be applied. But no consensus on such points can be assumed, and I shall therefore begin by distinguishing two different theoretical approaches whereby a scientific theory can be constructed.

One type of theory looks for possible ways of improving the status quo. It is guided by conceptions of perfection or health or optimum conditions in the broadest sense of the term. This was the line of thought pursued by Bacon and his followers. A scientific knowledge of natural principles and the avoidance of errors of judgment are not absolutely necessary for the preservation of the world, just as a knowledge of optics is not necessary for seeing properly. But they help to iron out flaws and gradually to improve the conditions in which people live.

The other type of theory is based on an assumption of improbability. Averse like the first type to the mere perpetuation of the status quo, it lays aside the routine expectations and certainties of everyday life and sets out to explain how relationships that are intrinsically improbable are none the less possible, and indeed can be expected to occur with a high degree of certainty. In contrast to Bacon, Hobbes based his political theory on such an assumption of improbability; and, unlike Galileo, Kant no longer relied on the possibility of an empirical knowledge of nature but cast doubt on synthetic knowledge as such and then investigated the preconditions for such knowledge. In this case, therefore, the major issue is not how to achieve practical improvements but how to answer a theoretical question that arises prior to any improvement, namely, how can an order be created that transforms the impossible into the possible and the improbable into the probable?

The following discussion remains strictly within the limits defined by the question just formulated, with the object of finding a suitable theoretical structure for the field of communication, as the only appropriate way of identifying the universal principles underlying all societies. But there are also practical motives making themselves felt with increasing urgency in a society geared toward growth and welfare. One can no longer proceed on the naïve assumption that improvements will always be possible on the basis of "nature," be it physical nature or human nature.[2] If nature is understood as improbability that has been surmounted, another standard has to be applied in assessing what has been achieved and what must be improved; at least then it becomes clear that the dissolution of an existing order implies a return to the improbability of a new combination.

Communication as a Problem

The type of communication theory I am trying to advise therefore starts from the premise that communication is improbable, despite the fact that we experience and practice it every day of our lives and would not exist without it. This improbability of which we have become unaware must first be understood, and to do so requires what might be described as a contra-phenomenological effort, viewing communication not as a phenomenon but as a problem; thus, instead of looking for the most appropriate concept to cover the facts, we must first ask how communication is possible at all.

Immediately, it becomes evident that a multitude of problems and obstacles have to be surmounted before communication can come about.

> The first improbability is that, given the separateness and individuality of human consciousness, one person can under-

stand what another means. Meaning can be understood only in context, and context for each individual consists primarily of what his own memory supplies.

The second improbability relates to the reaching of recipients. It is improbable that a communication should reach more persons than are present in a given situation. The problem is one of extension in space and time. The system of interaction among those present in each case assures, in practical terms, an adequate measure of attention for the purposes of communication, but the system collapses if a desire not to communicate is perceptibly communicated. Beyond the limits of this interactional system, however, the rules obtaining in that context can no longer be imposed. Hence, even if the communication finds means of conveyance that are mobile and constant over time, it is still improbable that it will command attention. In other situations people have other things to do.

The third improbability is the improbability of success. Even if a communication is understood, there can be no assurance of its being accepted. By "success" I mean that the recipient of the communication accepts the selective content of the communication (the information) as a premise of his own behavior, thus joining further selections to the primary selection and reinforcing its selectivity in the process. In this context, acceptance as a premise of one's own behavior can mean acting in accordance with corresponding directives but also processing experiences, thoughts, and other perceptions on the assumption that a certain piece of information is correct.

These improbabilities are not only obstacles preventing a communication from reaching its target; they also function as thresholds of discouragement and lead to abstention from communication if the prospects for it are thought to be inauspicious. The rule that it is impossible not to communicate applies only among those present within interactional systems,[3] and even then it merely states that communication will take place, and not what will be communicated. There will be a tendency to abstain from communication when the prospects of reaching people and of meeting with understanding and success seem to be poor. But without communication there can be no social systems. Hence, the improbabilities of the process of communication and the way in which they can be surmounted and changed into probabilities govern the formation of social systems. The process of sociocultural evolution can therefore be viewed as the transformation and expansion of the conditions for effective communication on which society constructs its social systems; this is clearly not just a process of growth but one of selection and of determining what kinds of social system are feasible and what kinds have to be rejected as too improbable.

The three types of improbability are mutually reinforcing. They cannot be dealt with and changed into probabilities one after another. The solution of one problem makes it that much more difficult to solve the others. The better one's understanding of a communication, the more grounds one has to reject it. When communication extends beyond the circle of those immediately present, understanding becomes more difficult and rejection again easier. The study of "philosophy" seems to owe its origins to this law of increasing mutual impediments.[4] When writing enables communication to extend beyond the audience present, limited in time and space, the rhapsodical element of rhythmical verse can no longer be relied on, since it can only carry with it the people actually listening; the subject matter itself must henceforth be the means of carrying conviction.[5]

This law that improbabilities mutually reinforce one another, and solutions to problems in one respect limit possibilities in other respects, implies that there is no direct way of achieving a progressive improvement in mutual understanding. Any efforts in this direction tend rather to run up against a growth problem coupled with increasingly irreconcilable demands. In the actual operation of the modern mass communication system, of course, people behave as though these problems have already been solved. In fact, they are no longer perceptible from the vantage point of particular offices in newspapers and broadcasting organizations. None the less, the question arises whether the structures of modern society are not essentially determined by the fact that the solutions to problems are mutually obstructive and generate a continual series of fresh problems.

The Concept of the Media

This theory requires a general concept covering the whole range of agencies involved in transforming improbable into probable communication in respect of all three basic problems. I propose to refer to such agencies as "media." Normally, we speak only of the mass media, a term applied to techniques—principally the press and broadcasting—used to extend communication to an absent public. Parsons has added the concept of symbolically generalized "media of interchange" and developed a corresponding theory on the analogy of money.[6] Since then the concept of the media has been used in two different senses in the social sciences and can only be understood from the particular context or with the aid of additional explanations. The suggestion that the concept be related to the problem of improbability in the process of communication and thus defined in functional terms might dissipate this confusion and at the same time help to clarify the significance and scope of three different kinds of media.

The medium that extends our understanding of communications beyond basic perception is language. It uses symbolic generalizations to replace, to represent, or to put together perceptions and to solve the resulting problems of mutual comprehension. In other words, language specializes in creating the impression of mutual understanding as the basis for further communication, however fragile the grounds for that impression may be.

The dissemination media are not adequately defined by the term "mass media." In particular, the invention of writing already fulfilled the function of transcending the bounds of immediate presence and face-to-face communication. Dissemination may be achieved through the medium of writing but also through the use of other procedures designed to preserve information in a fixed form. The selective influence of such media on culture can hardly be overestimated, since they enormously expand the store of memorized data available for additional communication, while at the same time restricting it through selectivity.

Generally speaking, communication theory has concentrated on these two types of media. But the resulting picture is seriously unbalanced. Only by endeavoring to discover which communication media are likely to be most successful can one develop a theory that really faces up to the problems of communication in society. The third kind of media may be described as symbolically generalized communication media because they alone effectively achieve the objective of communication.[7] With reference to social systems, Parsons mentions, as examples of this type of medium, money, power, influence, and value commitments. To this list I would add truth in the realm of science and love in the realm of intimate relationships.[8] The various media cover the major branches of the social system that have a civilizing influence and the main subsystems of modern society. This shows the extent to which, in the course of development, an increase in the possibilities of communication has been conducive to the formation of systems and the differentiation of special systems in the fields of economics, politics, religion, science, etc.

Symbolically generalized communication media can come into existence only when dissemination techniques enable the boundaries of face-to-face interaction to be transcended, and information to be stored up for an absent public of unknown proportions and for situations not yet exactly determined. In other words, they depend on the prior invention of a generally available form of writing.[9] In the face of such vastly expanded possibilities of communication, the guarantees of success provided by interactional systems, dependent as they are on physical presence, break down. They must be replaced or at least supplemented by more abstract and at the same time more specific means. Thus, in the Greek classical world, new code words *(nómos, alétheia, philía)* and correspondingly differentiated systems of standards were developed, denoting the conditions in which a probability of acceptance could still be assumed even though communication had become that much more improbable. Since then nobody has ever succeeded in combining all of the conditions for successful communication in a unified system of semantics applicable to all situations and, since the invention of printing, the differences between these communication media are becoming so pronounced that they ultimately break down even the premises of a unified natural, moral, and legal foundation to life: reasons of state and passionate love, methodically discovered scientific truth, money, and law all follow their different paths by specializing in different improbabilities of successful communication. They use different channels of communication—the state, for instance, uses the armed forces and the administrative hierarchy; passionate love uses the salon, the (publishable) letter, and the novel—and this leads to the differentiation of distinct functional systems, which ultimately make it possible to abandon an order of society based on fixed classes and allow modern society to take its place.

This brief sketch brings out the dual aspect of my theoretical concept. Order is created by virtue of the fact that communication, though improbable, is none the less made possible and becomes the normal situation in social systems. But the improbability of dissemination, once it has been surmounted by technological means, increases the improbability of success. New demands are made on culture as a result of changes in the field of communication technology. The established order of its media of persuasion comes under pressure from changed standards of plausibility, so that some elements become superfluous (for instance, the cult of the past) and others are encouraged (for instance, the cult of the "new"). All in all, a pronounced trend toward greater differentiation and specialization is discernible and hence also a need to institutionalize the arbitrary to an ever-increasing extent. At the same time, the pace of change is gradually accelerating, as generally happens in the course of human development,[10] so that means of overcoming increasing improbabilities in ever faster succession have to be developed out of what is already available, a task that becomes increasingly unrealistic if only on account of the time factor and leads to selection by the criterion of speed.

Modern Communication Facilities

Current discussions of the impact of the new mass media are restricted by their unduly narrow approach to the problem. Taking the concept of the "masses" as their starting point, they investigate the influence of the media on individual behavior. Viewed in this light, the social repercussions are due to the wholesale deformation of individual behavior by the popular press, films, and radio. Even changes only just taking shape in this sector, such as increased access to broadcast material, or indeed to communication in general within one's own home, are anticipated by reference to this point of view. I do not wish to deny the validity of this method of research. But when such a narrow approach is adopted certain important changes are entirely overlooked. For society must always be seen as a heterogeneous system; it does not consist merely of a large number of individual actions but is composed of subsystems and subsystems within subsystems, and it is only through association with such subsystems—for instance, the family, politics, economics, law, the health system, education—that actions can assume social relevance in the sense of repercussions being felt beyond the initial situation.

A much more comprehensive approach must therefore be adopted in order to gain a general picture of the changes being brought about in modern society because of the structure of its communication facilities. The problem of the improbability of communication in general and the idea of society as a heterogeneous system converge, since any system represents the transformation of the improbability of communication into the probable. Account must therefore be taken both of the changes in communication technology and of the different and changing prospects for successful communication as well as of the mutual repercussions of the two problem areas. In addition to all this, there is the question whether, independently of the medium, there may be, through the differentiation of systems, further direct effects on individual attitudes and motivation which, in the light of systems theory, appertain to the environment of the social system of society as a whole and react on it for this very reason. This problem of a latent, so to speak demographic, effect has recently made its way into analyses of the educational system as well, being reflected for instance in the catchphrase "hidden curriculum."[11] Similarly, it can be assumed (and in this context there are grounds for a comparison between the mass media and mass education in schools) that the organized mass media also operate selective restrictions on the repertoire of attitudes and motivations to which other subsystems of society can have recourse.

Of course, the scope of this article does not allow even an approximate description of such a wide-ranging program. I shall have to confine myself to a number of examples which may serve to illustrate some of the possible problems to be investigated.

However one defines the functional prerequisites for the preservation or development of a society, it cannot be assumed that the improvement in the prospects for successful communication will be equally advantageous to all functional spheres. The type of modern society that has its roots in Europe has hitherto been largely supported by a limited number of symbolically generalized communication media which have proved highly effective, more particularly by theoretically and methodically guaranteed scientific truth, by money, and by political power shared in accordance with the law. This reflects the prominence of science, economics, and politics in the general consciousness of this type of society. Even Parsons' theory of the general action system is based on the assumption that all functional domains can rely equally on a communication medium as a logical corollary of their differentiation. This is wishful thinking.[12] In any case, it will have to be accepted that there are neither natural nor theoretical guarantees for such a convergence of functional needs and communication prospects.

It is particularly noteworthy in this connection that no symbolically generalized communication medium has been developed to support the manifold activities designed to bring about change in individuals, ranging from education to therapeutic treatment and rehabilitation, although this is a functional domain totally dependent on communication. In this field, personal interaction remains the only way of convincing people of the desirability of change. Strictly speaking, there is as yet no scientifically reliable technology for this purpose.[13] Truth, money, law, power, love: none of these can offer adequate resources with sure prospects of success. An increasing amount of personal and interactional energy is being invested in this problem area without any real idea of whether or how technological inefficiency can be offset by such investment.

The above example shows that the problem of unbalanced development undoubtedly exists. In some fields the transformation of the improbable into what may be routinely expected is so successful that complex systems can be technologically controlled even though, in their basic processes, they depend on free decision making. In other fields development is at a standstill because, as performance demands increase, discouraging thresholds of improbability are reached even within simple interactional systems.

My next examples are drawn from an investigation of the repercussions of dissemination techniques on the functional divisions of society and on its communication media. The invention of printing clearly resulted in a very rapid transformation of the conditions in which important functions of the social system are fulfilled. Much of the development of religious radicalism that ultimately led to the splitting up of the various denominations was attributable to printing, because it publicly crystallized positions, making it difficult for their authors to retract them once they had been identified with them.[14] In the realm of politics, printing opened up opportunities for exerting political influence and making a political career outside court circles; renunciation of court office no longer necessarily implied renunciation of political influence,[15] and politics had to adapt itself to this new state of affairs. In the sphere of social life and intimate relationships, printing led on the one hand to increased educational opportunities and on the other to misguided aspirations; it was an incitement to imitation but at the same time exaggerated the possibilities of imitation.[16] It recommended rules but left their observance to the individual's discretion.[17] Generally speaking, therefore, printing changes the repertoires from which functional systems select their operations; it can broaden the range of possibilities but also complicate the process of selection.

This continues to apply when the mass media have become independent of education and have appreciably expanded their possibilities. But are there any identifiable guidelines? We can only resort to conjecture. A kind of media-based culture may develop whose sole justification lies in the fact that it is presupposed by the media programs themselves. But does this mean that morals corrupt power, as suggested by Arnold Gehlen with reference to the United States?[18] And are there not equally good grounds for the contrary assumption, namely that power can quite easily corrupt morals by changing the basic assumptions of the programs?

But there is less evidence to confirm such theories regarding mass media modification of basic political assumptions than to support the existence of more formal effects. Above all, the time structure of political action changes when it is constantly being reflected in the mass media. It tends to accelerate because politicians have to react from one moment to the next to the fact that, and the way in which, their actions are reported. The maneuvering that this entails effectively precludes consistent adherence to a political theory, and the conditions for participating in political life, though in one respect enormously expanded in democracies, are none the less restricted by the fact that it is necessary to keep constantly abreast of the latest developments.

However realistic such analyses may be, their starting point is the general assumption of the selectivity of all achievements in transforming the improbable into the probable. At each new and higher level of improbably probable communication achieved through improved technology, balance must be restored through new institutional expedients. And again, how can we be sure that satisfactory solutions will always be possible for each functional domain?

The problems discussed above with regard to the immediate repercussions of communication technology on functional systems must be differentiated from the question of whether the organized mass media system changes the personal attitudes and motivations to which society can refer for the purpose of encouraging socially acceptable behavior on a selective basis.[19] This, of course, has further indirect repercussions on the possibilities open to politics, science, the family, religion, etc. But these functional systems already exert a direct influence on the mass media without being pressurized by the motivations of their members. Take, for instance, the problems of church policy posed by the Küng case, in which provocation and reaction, courage and hesitation, reforming tendencies and conservative adherence to principles were all brought forward for the benefit of the mass media.

Leaving this aside, we may also have to consider the above-mentioned "demographic" impact of the mass media which consists of the formation of collective mentalities that subsequently give rise to conditions capable of affecting all social systems. But this certainly does not warrant the conclusion that uniform, mass attitudes are generated among the population in this fashion, for instance by television. It is more realistic to assume that certain principles followed in determining whether something should be printed or broadcast are passed on to the public; and it is in fact such principles that define what shall appear as information.[20] Perhaps the most important principle of this kind is that a thing should seem new or out of the ordinary in order to be worth reporting. This does not rule out, but rather includes, monotonous repetition (football, accidents, government communiqués, crime). Another similar principle of selection is conflict.[21] It must be assumed that such principles, which constantly stress discontinuity as opposed to continuity, tend to undermine confidence. It is quite conceivable that they stimulate simultaneous demands for protection against and participation in change, thus generating both fears and claims. Society's political and economic system, whether it is held together by a private capitalist or state capitalist order, may thus find it increasingly difficult to meet the expectations of the population.

"Are we asking the right questions?" was a concern voiced at a Unesco conference on the mass media.[22] And even at the end of my outline of problems we still cannot be sure whether the questions being asked are the "right" ones, while a philosopher will be inclined to ask whether the "right" questions exist at all. None the less, it should be possible to develop a more radical and systematic approach to the study and solution of problems in the field of communications research than has hitherto been the rule. The connection between improbability and the formation of systems is one of the concepts that systems theory has to offer in this context. If the problem of improbability is taken as the starting point, there is an automatic tendency to ask if not the right questions at least more fundamental ones that recognize that the issue of the connection between communication and society is not confined to the field of communications research but is in fact central to all social theory.

Endnotes

1. In *Kommunikation: Ein Begriffs- und Prozessanalyse* (Opladen: 1977), Klaus Merten attempted to analyze such discussions with a view to identifying common characteristics.
2. For statements of this kind see, for example, Joseph Glanvill, *The Vanity of Dogmatizing* (London: 1661); Francis Hutcheson, Preface to *An Essay on the Nature and Conduct of the Passions and Affections* (London: 1728).
3. Paul Watzlawick, Janet H. Beavin, and Don D. Jackson, *Pragmatics of Human Communication: A Study of Interactional Patterns, Pathologies and Paradoxes* (New York: 1967), pp. 48, 72 et seq.
4. See Eric A. Havelock, Preface to *Plato* (Cambridge, Mass.: 1963).
5. On the development of nonverse literary art forms, ses also Rudolf Kassel, "Dichtkunst und Versifikation bei den Griechen," lecture to the Rheinisch-Westfälischen Akademie der Wissenschaften, 1980.
6. The most important essays on this subject have been recently reprinted in Talcott Parsons, *Politics and Social Structure* (New York: 1969). See also Talcott Parsons, "Social Structure and the Symbolic Media of Interchange," in Peter M. Blau, ed., *Approaches to the Study of Social Structure* (New York: 1975), pp. 94–120. Noteworthy among the numerous secondary commentaries are: David A. Baldwin, "Money and Power," *The Journal of Politics* (1971), no. 33, pp. 578–614; Rainer C. Baum, "On Societal Media Dynamics," in Jan J. Loubser et al., eds., *Explorations in General Theory in Social Science: Essays in Honor of Talcott Parsons* (New York: 1976), 2:579–608; Jürgen Habermas, "Handlung und System — Bemerkungen zu Parsons' Medientheorie," in Wolfgang Schluchter, ed., *Verhalten, Handeln und System — Talcott Parsons' Beitrag zur Entwicklung der Sozialwissenschaften* (Frankfurt: 1980), pp. 68–105; Stefan Jensen and Jens Naumann, "Commitments — 'Medienkomponente einer ökonomischen Kulturtheorie?" *Zeitschrift für Soziologie* (1980), no. 9, pp. 79–99; and Stefan Jensen's Introduction to his edition of Talcott Parsons' *Zur Theorie der sozialen Interaktionsmedien* (Opladen: 1980).
7. Although the issue is adequately understood from the point of view of content, the question of terminology is still wide open. Following Parsons, some use the term "exchange media," some "interactional media" and some "communication media," None of these is quite satisfactory. As is often found in the case of new theoretical discoveries, our existing vocabulary provides no exactly suitable term.
8. See Niklas Luhmann, "Einführende Bemerkungen zu einer Theorie symbolisch generalisierter Kommunikationsmedien," *Soziologische Aufklärung*, 2:170–192 (Opladen: 1975); and, on Parsons' theories, Niklas Luhmann, "Generalized Media and the Problem of Contingency," in Jan J. Loubster et al., *Explorations in General Theory*, pp. 507–532.
9. For developments in the Greek *polis* that are of decisive importance in this context, see Jack Goody and Ian Watt, "The Consequences of Literacy," *Comparative Studies in Society and History* (1963), no. 5, pp. 304–345.
10. See Gerard Piel, *The Acceleration of History* (New York: 1972).
11. See, in particular, Robert Dreeben, *On What Is Learned in School* (Reading, Mass.: 1968), and its probably, on the whole, unduly optimistic assessment.

12. Thus, critics have noted the inherent limitations of an analogy between money and other communications media. For a recent discussion of this subject see, in particular, Habermas, "Handlung und System."

13. See Robert Dreeben, *The Nature of Teaching: Schools and the Work of Teachers* (Glenview, Ill.: 1970) in particular pp. 26, 81, 82 et seq.; Niklas Luhmann and Karl Eberhard Schorr, "Das Technologiedefizit der Erziehung und die Pädagogik," *Zeitschrift für Pädagogik* (1979), no. 25, pp. 345–365.

14. See Elisabeth L. Eisenstein, "L'avènement de l'imprimerie et la Reforme: une nouvelle approche au problème du démembrement de la chrétienté occidentale," *Annales ESC* (1971), no. 26, pp. 1355–1382.

15. On this topic I would recommend J. H. Hexter's *The Vision of Politics on the Eve of the Reformation: More, Machiavelli and Seyssel* (London: 1973).

16. An issue that has been much debated since the seventeenth century, especially with reference to women. See, for example, Jacques du Bosq, *L'honneste femme*, new edition (Rouen: 1639), especially pp. 17 et seq.; Pierre Daniel Huet, *Traité de l'origine des romans* (Paris: 1670; reprinted Stuttgart: 1966), pp. 92 et seq. For a modern view, see also Georg Jäger, *Empfindsamkeit und Roman* (Stuttgart: 1969), pp. 57 et seq.

17. See Erich Köhler, "Je ne sais quoi: Ein Kapitel aus der Begriffsgeschichte des Unbegreiflichen, *Esprit und arkadische Freiheit: Aufsätze aus der Romania* (Frankfurt: 1966), pp. 230–286; Christoph Strosetzki, *Konversation: Ein Kapitel gesellschaftlicher und literarischer Pragmatik im Frankreich des 18. Jahrhunderts* (Frankfurt: 1978), especially pp. 125 et seq.

18. See Arnold Gehlen, "Die gewaltlose Lenkung," in Oskar Schatz, ed., *Die elektronische Revolution: Wie gefährlich sind die Massenmedien?* (Graz: 1975), pp. 49–64.

19. On the underlying theoretical concept see Niklas Luhmann, "Interpenetration: Zum Verhältnis personaler und sozialer Systeme," *Zeitschrift für Soziologie* (1977), no. 6, pp. 62–76.

20. Here I am assuming an information concept whereby something can be regarded as information only if it is selected according to the criterion of difference. This means in turn that a comparative model is assumed for the purposes of identifying information, but this is not simultaneously conveyed to the public and thus cannot (or can only with difficulty) be controlled by or elicit a communicative reaction from the recipients.

21. On this subject, see, in particular, Hans Mathias Kepplinger, *Realkultur und Medienkultur: Literarische Karrieren in der Bundesrepublik* (Freiburg: 1975).

22. *Mass Media in Society: The Need of Research* (Paris: Unesco, 1970) (Reports and Papers on Mass Communication, 59).

5.
Modes of Communication and Society

At no time has society been able to foresee or even to observe deep structural changes. The invention of writing, the invention of the alphabet, the invention of printing passed almost unobserved. Certainly, contemporaries were not able to assess the importance of these events, nor were they able to foresee their consequences in terms of a structural revolution of the whole societal system. Take the alphabet. Poets looked for an improvement of their mnemotechnic devices and had to change the traditional code of writing to be able to transcribe the complete content of their oral texts. If daring, they aspired to become more or less independent of the muses and thereby provoked a religious scandal—Simonides of Keos being the famous example. They never did know that they set off a complete change in the ways humans relate to the world. In fact, the alphabet made it possible to write for readers and to presuppose a common distance from objects, a critical attitude, and, thereby, new dispositions to change the circumstances of human life.

This difficulty—and this may disappoint your expectations—is not due to the lack of social science and research capacities in the past. It is not an obstacle that may be surmounted, today or tomorrow, by improved empirical and analytical skills. It is not simply a question of more knowledge. There are structural reasons that seem to limit our possibilities, and our wisdom may well have to remain at the level of Plato's sophrosyne: to know what we know and what we don't know.

We have to distinguish at least two kinds of problems that seem to accumulate:

First, it is generally difficult to observe events. It is much more demanding than observing states. To mark an event as decisive, you have to identify both the before and the thereafter and to know that it is a particular event that makes the difference.

And second, changes in the media and techniques of communication are not marginal improvements. The system of society consists of communications. There are no other elements, no further substance but communications. The society is not built out of human bodies and minds. It is simply a network of communication. Therefore, if media and techniques of communication change, if the facilities and sensitivities of expression change, if codes change from oral to written communication, and, above all, if the capacities of reproduction and storage increase, new structures become possible, and eventually necessary, to cope with new complexities.

However, these changes *in* communication have to be introduced *by* communication. There is nobody outside the system who could plan and direct it. The system evolves by self-reference. It can be managed and controlled only by parts of the same system, i.e., by itself. Observing and describing, planning and directing the system presupposes the system, and not only as object but also as subject of its own activities. Therefore, most important steps in evolution had to be taken before the final supporting context could be developed, namely with support by a provisional function which later could be dropped. Language probably evolved by using signs. It cannot have been from the beginning a structure of mental and communicative operations. Money evolved as a means for solving accountancy problems in large households, and not as a medium of exchange. And the code of lovemaking has been invented for pleasure, not for marriage.

Given these typical structures of the unplanned evolution of self-referential systems, how could we proceed in assessing the importance and the consequences of what seems to be rather radical changes in present-day communication technology? Most literature, obviously, concentrates on minor problems that attract attention during the process of change. Maybe television will prevent children from reading. Maybe computers will lead to unemployment within the more traditional trades. Maybe the heroes of our future will be computer-addicted young men—similar to the boy in the movie *War Games*. Maybe computer mentality, focusing on a certain technique of decomposing problems, will cut out all sensitivity for the deeper mysteries of life. We may tend to perceive what we lose. But we are unable to observe and describe the system of society in the process of a structural change. And of course, we cannot plan and we can not prevent change at this level.

Being aware of these fundamental problems of self-observation and self-description of the societal system does not lead to resignation. On the contrary, it is a prerequisite to any realistic approach. We cannot observe and describe the future society, but we may be able to see what kind of structural change is going on. We may not be able to fix the event between the before and the thereafter. But we can at least see in which respect fundamental limitations of previous communication seem to change.

The difficulty, however, is a typical problem of complexity. We live within *one* society, *one world society of social communication*, but we have to cope with *several structural changes at once*. This, in addition, makes prognostication impossible because we would have to account for the future effects of an interference of different kinds of structural change.

1. The first type of change can be described as *increasing control capacity*. But beware: we have to use the German and not the English sense of the term, and this distinction is decisive. English-speaking people think of control as some kind of steering or even domination. It is based on the *capacity to compare (i.e., check on) input to output*, or *input to goals*. The increase of this capacity would increase our ability to reach our goals even under conditions of increasing complexity. It seems to be very doubtful that this is, in fact, the case.

In its narrower sense, control means *comparison of input to memory*, to memory only; that is, comparison of the present not to the future but to the past. Control in this sense means looking backward. This capacity has indeed been extended continuously since the invention of writing, and of course printing. The computer technology again multiplies our storage capacity, our capacity to handle and analyze stored information and to compare it to incoming news. It does not at all improve our capacity to reach our goals. Goals may, of course, be part of our memory. In this sense, they are given as *past fancy*. Thus, increased control technology may mean increased capacity to perceive—disappointments. The immediate result will be the experience of more, not less, disappointments. And in fact, everybody today talks about "crisis" and, as we all know, crises can be talked into becoming in fact unsolvable problems. Increased storage capacity must mean that we fall more and more into the dead hand of the past—of past facts and past fancies.

Our religion is prescribed by a book. Our political hopes are prescribed by the slogans of the bourgeois revolution two hundred years ago. Our conflicts are structured by the distinction of capital and labor, stored in books and ideologies, organizations, and even political systems, although we could know that no real problem of our society can be solved by shifting money from capital to labor or back.

I won't dispute the fact that we continue to invent new things, new technologies, new possibilities of communication and that, on a technological level, we may well be able to see the near future as bringing desirable improvements, higher capacities, improved methods of production at diminishing costs. But the fascination of newness itself depends upon control capacity. New is something in comparison with the past, and the fashion of admiring newness and originality since the sixteenth century may well have been a result of the printing press. The point is that, as far as control capacity is concerned, the search for improvements will at the same time increase the power of our memory, the power of our past. We may become unable to forget.

2. Increased control capacity, however, is only one of the visible changes. Another and possible equally important point refers to *new possibilities to communicate movement*. Until very recent times we had to choose between oral communication and writing (including print). This was a sharp difference, very important in its consequences. Now we have movies and television. Oral communication too has been expanded considerably by the telephone and it will expand further in the near future by new kinds of—may I say "electronic orality."

These developments change the relation of communication and time. Formerly, only the communicative act had to be time related, and communication itself, by necessity, had a time-binding function (to use the terminology of Alfred Korzybski). Now, the *content* of communication may become time dependent—and this not only by complicated circumscriptions, but directly. Time becomes visible—and this not only in nature and not only by perception, but as a result of communication. We become able to see vanishing events, movements, and changes as a result of prepared and carefully selected communication. How can we control the selectivity, the possibility of error, the possibility of deception?

By television in particular we can communicate synchronized visible and audible events. This has never been possible before. But then, what does it mean to *communicate* it? If now *everything* becomes a possible object of communication, bypassing the circumstantialities of language, and if *nothing* remains exempt, communication may well lose its own specific function to add something to the world. It becomes, more than ever before, a reconstruction of the world and, thereby, a source of disappointments. And again, how could we impose corresponding criteria for selection and responsibilities if the whole process has a totalizing circularity?

This is, of course, a very general problem, and it has ancient roots. Language, and above all writing and printing, require a semantic apparatus of protection against unintentional and intentional misuse of symbols. At the very center of our culture we have created codes that fulfill this function: codes for truth and for legitimate power (potestas), codes for paying with money, and codes for real, sincere love protecting against mere seduction. In fact, alphabetic writing stimulated the articulation of these generalized media of communication which recombine selection and motivation. These codes were invented as, and continue to operate as, social devices of protection against the dangers of language. But will they work as effectively against the seduction of moving pictures? And how could, in this case, the intention behind the communication be perceived, grasped, attributed—not to speak of controlled? On a rather superficial level, we are well aware of this new power of persuasion and seduction which is no longer the power of language only. For its control, however, we rely, more or less, on organizational techniques, supervising agencies, political bodies, commissions for discussing and framing principles of professional ethics. We have nothing equivalent to the silent efficiency of what once seemed to be sufficient: truth. There may be quite a new power of creating convictions simply by flashing immediately disappearing events on the mind, which may then, apparently by itself, draw its conclusions.

3. My third point partly combines the previous ones and partly adds something new. One of the most spectacular aspects of recent developments seems to consist of new possibilities of combining different communication technologies by using the computer. Within a short span of time we may be able to "write"—could we then say "write"—new books by looking at the computer for existing knowledge, talking to it (of course not simply in our best BBC English but with a new kind of carefully learned electronic speech), and looking again at the results for necessary corrections. We may no longer need a publisher, but perhaps an experienced book production assistant. To print out or to leave it in the computer for further

writing may be a matter of choice. Whoever will use the never-finished book—could I say "the book"?—may ask for the permission to print it for himself. Or: Do we need to go to the confessional any more—talking to the priest who himself gives advice and absolution on the basis of his manuals of moral casuistry, be it the Summa Angelica (of Angelus de Clavisio), the Sylvestrina (of Sylvester Prierias Mazzolini), or more recent treatises *de casibus conscientiae*. All of these rules may be available as software so that we will be able to work off our sins with our home computer. Something less practical may remain the preferred way to do it—but then, this will survive only by the grace of this preference.

This may be enough of science fiction, just to show the possible impact of these new modes of combining different forms of communication, transcending the trenchant difference of talking and writing. My last remarks are a further warning with respect to oversimplified theories of structural change. We are hardly able to predict the coming state of our society—its total destruction being the only thing that we know for sure to be possible. But in addition, we also have problems in attributing change to a change in communication technologies. Obviously, there are other sources of change to be taken into account. No sociologist, at present, could give you a complete overview of such sources. Again an example may suffice to make the point.

Looking at the history of literary fiction we can observe a curious kind of change, concerning the complexity of characters. (I don't speak of the complexity of plots!) Until the rise of the novel in the eighteenth century, the characters had to be heroes or villains, that is, "heavy" characters, and a stereotype would be sufficient to identify them. Apparently, surviving restrictions on oral textuality were at work. One had to be able to characterize them in highly formulaic oral speech: She was a beautiful young widow, belonging to one of the most excellent families of the country. This would do the job. Next came a time, beginning perhaps with Richardson, when characters had to be complex, and eventually open to diverging interpretation and interesting to different readers, never quite finished, self-reflective, and never quite sure about themselves. Today, we seem to prefer again what E. M. Forster[1] calls "flat" characters, as opposed to "round" characters: the good boys of Walt Disney or the characters of soap operas, the characters of television series or the characters of their commercials. We can parody them. Then they come out as Superman, Spiderman, Batman, and Wonderwoman, who seem to be enjoyable even when we fail to notice the parody. We seem to have come full circle, to have arrived again at a reduced simplicity—just the minimum of information to get the story across. However, can we really explain this change from heavy to round and from round to flat characters by the sequence of orality, writing/printing, and finally television as the dominant mode of communication? Some of these flat characters came, by the way, from the comics, i.e., from print, to the movies. How can we know and how can we prove that communication technology is in fact the decisive variable? Although a very suggestive hypothesis at first sight, this might not outlast a second look.

Above all, we have to take into account the fact that literature reacts upon literature and that the fashion of complicated characters wears thin after some time. Imitation of "real life" may have to be replaced by irony or by real "real life," which is not that complex but rather trivial and simple. The story has to hit a problem, but character, after all, may be a minor one.

Texts may adapt to changes in communication technology; they may adapt to the previous history of producing texts and to the cultural imperative to appear as new. They may also adapt to changing fashions and, of course, to an increasing or decreasing range of the public that enjoys them. There are many influences that may reinforce or counterbalance each other. One-factor theories can give only highly simplified versions of our society. They may be misleading or outright wrong. They may be nevertheless successful if reintroducing the description of communication into the system of communication requires this kind of simplification.

The historical-comparative approach I did pursue in this essay does not focus on causes and consequences, but on problems. There is:

> *first*, an increasing gap between control capacities and goal attainment which leads, among other things, to increasing disappointments and negative feelings toward society. There is
>
> *second*, a quite new representation of reality-in-movement, the impact of which is difficult to appreciate. Everything now is a possible object of communication, but communication may not be the same thing as before. And
>
> *third*, there is a quite new ease in combining different communication technologies, changing from situation to situation according to needs and chances. This new elasticity may solve or

attenuate many problems, but we really don't know what kind of semantics, what kind of culture will be appropriate for it, replacing our old culture, which had developed in reaction to writing.

Endnote

1. E. M. Forster, *Aspects of the Novel* (1927; reprint London: Harcourt, 1949). I refer also to a paper of Christine Brooke-Rose, "The Dissolution of Character in the Novel," given at a conference on the "Reconstruction of Individualism," Stanford University, Stanford, California, February 1984.

6.
The Individuality of the Individual: Historical Meanings and Contemporary Problems

Reconstructing individualism is not an easy task. It is not easy because it has been tried before. The question of how to conceive individuality has a long scholastic tradition, characterized by what we now tend to see as outmoded sophistication. Even if we concentrate on the human individual, we must travel back at least two hundred years if we want to survey the full array of theories and discover why they failed or lost their power of persuasion.

It would be unwise to begin without casting at least a superficial glance at what has already been done. Sociological theory, however, will not prove very helpful. It conceives history as a process of increasing individualism. Its classics contain an important theory built around this point: increasing social differentiation leads to increasingly generalized symbolic frameworks, which make it increasingly necessary to respecify situations, roles, and activities, which results in increasingly individual human beings. Traditional societies had to restrict the number of possible role combinations because families were located unambiguously in a hierarchical system of status groups. Modern society no longer situates families in this manner. Stratification has lost its former importance, and, as a consequence, the choice of role combinations must be left to the individual.

I am indebted to Stephen Holmes for reading and improving an earlier draft of this essay. The essay originally appeared in Thomas C. Heller, Morton Sosna, and David E. Wellerby, eds., *Reconstructing Individualism: Autonomy, Individuality, and the Self in Western Thought* (Stanford: Stanford University Press, 1986).

This is one of two currents of mainstream sociology, represented at its best by Durkheim. Within the other tradition, stemming from Simmel and Mead, the individual is conceived as an emerging unit —emerging not from history but from social encounters. Simmel sees the identity of an individual as a collage, glued together by the viewpoints and expectations of other individuals. The fragmented self of one's own self-impressions becomes continuous and reliable only in and through social situations. Similarly, Mead thinks of the individual mind as an emerging unit, an inner copy of social interaction.[1] Here we have the beginnings of a social theory of the individual, a theory that decomposes and recomposes the individual with reference to social conditions. The social, however, is seen solely as interaction, not as society, and as a more or less present occasion, not as history.

Sociology has reacted unsympathetically to the ideological antithesis between individualism and collectivism (or between the individual and society), trying to bridge the gap and yet give each side its due. It has not taken up the difficult task of formulating a theoretically relevant concept of individuality. Fortunately, we do not depend upon sociology alone. European intellectual history contains a long series of attempts to define and promote individuality. For our purposes, this history provides a double perspective. First, it serves as a pool of intellectual resources and, above all, of warnings: there are many traps on the way to a theory of the individual, traps that we can avoid if we know about them in advance. And second, it allows us to reconstruct the history of reconstructing individualism in the style of the sociology of knowledge, relating the history of ideas to that of social structure. The first perspective, *historia magistra vitae* (history teaches life), may be sociologically naive, but we should not judge it prematurely.[2] The second reminds us that every reconstruction of individualism must be carried out within society, by people who think of themselves as individuals and set out, so to speak, to reconstruct themselves.[3]

By Descartes' time, medieval scholastic debate had settled one thing about the individuality of the individual: individuality cannot be defined by pointing to some special quality of the individual in counterdistinction to other qualities; it is not something given to an individual from the outside. An individual is itself the source of its own individuality; the concept of individuality therefore has to be defined by self-reference.[4] All kinds of individual beings, not only humans, are defined by self-reference.[5] The scholastic theory of individuality was, of course, written by and for humans who had to make up their minds about themselves and their social conditions. Thus, the special importance of self-reference for defining the human individual is not surprising. Amid religious schism, political wars, emerging sovereign states, and economic progress and decline, self-reference, which reconstructs the individual on the basis of its own problems and resources, must have seemed an attractive refuge.[6]

One of the most interesting results was the devotional movement of the seventeenth century, which privatized the attempt to achieve salvation. On the basis of a religious world view, the movement fought against a competing tendency to associate individuality with libertinage, with a *fort esprit* that defied religion. The difference between salvation and damnation remained decisive.[7] But religious care was no longer care for others. It did not require praying for others, monastic conditions, or supererogatory works.[8] Instead, it was care for one's own sole salvation. "Etre devot, c'est vouloir se sauver et ne rien negliger pour cela" (to be devoted means to care for one's own salvation and to neglect nothing in view of this aim)—that is the Jesuit view of the job, and it seems compatible with, even similar to, other jobs.[9]

Today, such self-centeredness is no longer à la mode, but at least two effects of the devotional movement should be kept in mind. First, except for alms and charity, necessary for his own salvation, the individual's orientation toward the experience of others was remarkably devalued. The social self, the "me" of James and Mead, received bad marks like "pride" and "vanity." If you took the role of the other, admiring your devotion, you were already on the wrong track: devotion cannot be communicated, at least not intentionally.[10] This led to a second recognition: real and false devotion become indistinguishable. Sincerity and authenticity cannot be communicated; but if others cannot know his sincerity, the individual will feel unable to trust himself.[11] The same problem arises in love affairs.[12] Whoever tries to convince another of his love becomes insincere by attempting to do so. The only escape seems to be a profession of insincerity.[13]

To sum up: within the devotional movement, self-reference—the very essence of individuality—was judged to be the ruin of individuality, at least if an individual tried to stay on the path to salvation and professed to do so. Salvation presupposes something to be saved, whereas an individual, in the literal sense, is a being with an indivisible, indestructible soul. If we drop this religious or ontological warrant of individuality, we arrive at the *homme aimable*, the sociable person of the eighteenth century. He too has lived his life, and he too has left a clear record of his experiences.

In the eighteenth century, two resources alone remained available for reconstructing the individual as sociable person: the reevaluation of nature and the reevaluation of sociality. Thus a new cult of sensitivity and friendship replaced religion; self-love expanded to include concern for others. (We are approaching Simmel and Mead.) Symbolic interactionism was tested for the first time during the eighteenth century. Social interaction requires superficial conversation without risks or consequences. Its essence is to take the role of the other and to avoid bothering others with one's own problems or peculiarities; not to speak about oneself is one of its central norms.[14] The results were disappointing, particularly for the self-conscious individual: somehow, the individual withdrew from interaction. By the end of the eighteenth century, the *homme du monde*, the *homme de bonne campagnie*, was no longer an individual. If he dies, observes Sénac de Meilhan, what do we know about him? That he had a seat in the Opera, that he liked to play lotto, that he took his supper in the city.[15]

Given that this was an age of *lumières et dégoûts* (enlightenment and disgust),[16] how could the individual maintain his individuality? Two ways were open to him. Each required a new distinction, a new guiding difference (in German I could say *Leitdifferenz*), which replaced the difference between salvation and damnation. In the first, the individual was guided by the difference between nature and civilization. Since there was no way back to nature, men must come to terms with lost innocence and with property. Thus conceived, the social order retained and even reinforced some of the bad characteristics of false devotion, such as pride and vanity. In it, self-interest was given a civilized disguise. It was an order of hypocrisy and, above all, of visible injustice—in which, for example, property was distributed unequally. In the decades before the French Revolution, social theories characterized civil society (in contrast to the state of nature) in terms of the difference between property and indigence and of a law of necessarily unequal distribution.[17] The individual might identify with opulence or with misery, according to his situation. Labor became the only link connecting these extremes, and social analysis showed (as a kind of comfort?) that both the rich and the poor were slaves of civil society. They were not really individuals. They did not really belong to themselves.[18]

Fortunately, property was only one way to reconstruct the individual within a framework of social interdependencies. The other way used the concept of art and, later, the theory of aesthetics. Nature was said to produce, from time to time and in exemplary individuals, what came to be called "genius."[19] The social was introduced with concepts such as "taste" *(gusto, goût)* and "the public," which judges not by reason but by heart and sentiment. The self-reference of an individual is the self-reference of his heart, and the work of art is, like nature itself, an outside condition that stimulates the heart to act upon itself.[20]

The problem is that all of this depended on social stratification. Taste was a qualified taste, and the public was a restricted public.[21] Swiss and German authors (Crousaz, Bodmer, Gottsched, Baumgarten, Kant) tried to eliminate taste as the ultimate judge, replacing it by a new guiding distinction: the difference between the particular and the general. They tried to find, within the realm of concrete particularities, general criteria for the beautiful. If a rational proof for such criteria could be provided, it would be possible to desocialize the judgment of individuals. No longer would the public (of connoisseurs) define the criteria (without reason), but the criteria would define the competence of the public. In making a judgment, the individual would no longer depend on his social stratum but on the realization of the general within himself.

Only the results of the long and complicated dispute over general criteria for judgment and taste are of interest here.[22] They lead to a reconstruction of the individuality of the individual at a new intellectual level. After Kant, a new kind of subjective individualism became possible; given the turn to the "transcendental," the facts of consciousness had to be evaluated by a kind of double standard: empirical and transcendental. This difference parallels that between the particular and the general. As a result, the individual (not only the Cartesian mind) emerged as the subject, as subject of the world. Experiencing the world, the individual could claim to have a transcendental source of certainty within himself. He could set out to realize himself by realizing the world within himself. Education, which bourgeois theory saw as suffocating the voice of nature,[23] now became a liberating process, for which the French Revolution had changed the scene.[24]

The individual as the subject? This cannot be surpassed. We can only forget what it meant and then try to salvage one or another of its secondary meanings—say, emancipation, reasonable self-determination, or understanding of the subjective point of view in everyday life.

The history of the individuality of the individual does not continue beyond this point. Or rather, it continues only as a history of ideologies, as a history of "individualism," a term invented in the 1820s.[25] I hope that the title of this conference on "reconstructing individualism," despite its formulation, is not meant to suggest a reshuffling of ideologies. So I shall stop my historical report here and return to a theoretical analysis of the individuality of the individual.

Within the general framework of the theory of society—society being conceived as the encompassing social system—I propose to characterize modern society as a functionally differentiated social system.[26] The evolution of this highly improbable social order required replacing stratification with functional differentiation as the main principle of forming subsystems within the overall system of society. In stratified societies, the human individual was regularly placed in only one subsystem. Social status *(condition, qualité, état)* was the most stable characteristic of an individual's personality. This is no longer possible for a society differentiated with respect to functions such as politics, economy, intimate relations, religion, sciences, and education. Nobody can live in only one of these systems. But if the individual cannot live in "his" social system, where else can he live? As *homo viator* on his way to heaven or hell?

I would suggest that the change from stratification to functional differentiation as the basic principle for subdividing society explains the successive historical attempts to reformulate the basic problems of individuality. "Explains," of course, is a strong term, but we can discern a remarkable correspondence between structural and semantic changes.[27]

At first, the increasing pressure to live an individual life might have led, in Protestant and Catholic churches alike, to a privatizing of the path to salvation. This ended by increasingly differentiating religion from other social domains. At court and in high society, to be religious or not, according to prevailing opinion, came to be a fashion, even to be described as fashion. This made devotion visible and identified it as a social phenomenon; thus it ruined the devotional movement that had flourished during the seventeenth century by opening to question two of its major tenets. The alliance between the court and religion was no longer structurally important, but a matter of the day, of caprice, of being à la mode.

As a sociable person, the individual could survive, finding in his present life not heaven or hell, but pleasure or ennui. This solution, however, was in turn undone by new differentiation. Society, transforming itself into a functionally differentiated system, could no longer be represented on the level of social interaction, not even by or within its highest strata. The interaction of high society, still representing the "good society" within the societal system, became an island of social rationality, isolated from all serious business. In consequence, the difference between pleasure and ennui collapsed, making a meaningful life impossible. The depersonalization of social interaction desocialized the individual.

There is, after all, a hidden relation between the functional differentiation of the societal system and the individual's self-proclamation as subject. Given the traditional connotations of *hypokeimenon/subiectum*—something "lying under" and supporting attributes—"subject" means something that underlies and carries the world, and, therefore, something that exists in its own right as a transcendental and not as an empirical phenomenon. The guiding difference is no longer pleasure and ennui, or self and other; it is the difference between empirical particularities and transcendental generality. The individual leaves the world in order to look at it. I interpret this extramundane position of the transcendental subject as a symbol for the new position of the empirical individual in relation to a system of functional subsystems. He does not belong to any one of them in particular but depends on their interdependence. But is this a good symbol? Is it an adequate semantic correlate of functional differentiation? What about its traditional connotations, its exaggerations, its internal plausibility, its ideological, political, and motivational feedback?

Today the subject is again fashionable, at least in sociology. Partisans of the subject attack behaviorism, systems theory, information technology, and survey research, and ask for the recognition of the subject. They no longer understand, however, what the word means. They recite outworn philosophical terminology in an erroneous way. We should drop the term "subject" ("psychic system," "consciousness," "personal system," perhaps even "individual" would do the job) if we are simply referring to a part of reality. How can we conceive of a part of reality as underlying or supporting reality? Can a part simultaneously be the base of the whole? Can the subject be the *subiectum* of itself and of everything else? Can we conceive the subject as *sphaera se ipsam et omnia continens* (as containing itself and everything else)? This, of course, was the old definition of the world.[28] Can we conceive of the subject as a duplication of the world? The transcendental theory of consciousness was at least aware of this problem and tried to solve it by claiming extramundane status for self-referential conscious experience. But if we refuse to accept this transcendental solution, and of course we do, we are again faced with the old paradox of privileged parts supporting a whole.

For all of these reasons, reconstructing individualism today cannot mean reaffirming the subject. We should honor the subject by finding it adequate successors, adequate both in terms of the problem and in relation to the social structure of contemporary society. To make concrete proposals, of course, is risky and difficult. Nevertheless, we can begin with a simple but far-reaching observation concerning the recent boom in research on what have come to be called "self-referential systems." Summing up what has been done and anticipating what can still be done, we can characterize this movement in a few statements:

1. Self-referential systems are empirical and have no transcendental status whatsoever. They are normal objects of normal science, but recognizing their existence has important epistemological consequences.

2. There are several types of self-referential systems, which differ according to their basic operation. This can be life (or possibly some other, or even any, material or energetic operation), consciousness, or communication. Mixing such operations is impossible, because they presuppose closed systems. Conscious systems are not living systems, social systems are not conscious systems. The different domains may, of course, be causally interconnected, yet they are not simply relations between facts but are always organized as a relation of a system to its environment. In this sense, living systems and conscious systems are parts of the environment of social systems, whereas social systems are part of the environment of living systems and conscious systems.

3. The term "self-reference" refers not only to the identity of the system (as does "reflection" in its classic sense, e.g., as used in the philosophy of consciousness of German idealism) and not only to the structure of the system, that is, to its morphogenesis and its self-organization. It refers also, and primarily, to the constitution

of its basic elements. The elementary units of self-referential systems can be produced and reproduced only as self-referential units. They combine self-identity and self-diversity (to use Whitehead's formulation),[29] and their status and function as elements mean that this combination cannot be dissolved by the system itself. This means also that the system cannot distinguish its own basic elements from its own operations.

4. Such systems are called, following Maturana, "autopoietic."[30] Autopoietic systems are closed systems in the sense that they cannot receive their elements from their environment but produce them by selective arrangement. I could also say—and this reminds me that I am operating very close to my background in transcendental theories—that everything used as a unit by the system, whether its elements, its processes, or the system itself, has to be constituted by the system. Units, of course, are complex facts and can be analyzed by an observer. The system itself, however, has to rely on self-constituted reductions to link and reproduce its own operations. The unity of the system, therefore, is nothing but the autopoietic process of constituting units for itself within itself.

5. Producing units requires reducing complexity, namely, the complexity of a domain in which distinguishing between system and environment makes no sense. The autopoietic system has to use distinctions and indications as basic operations (in the sense of Spencer Brown's logic).[31] It realizes the closure of its own operation by self-indication, but to identify itself it needs to distinguish system from environment. It does not necessarily need knowledge about its environment, but it needs to be distinct from that environment.

6. Autopoietic systems produce their elements within temporal boundaries, depending on a beforehand and a thereafter. Many of them, certainly conscious systems and social systems, consist of events only, of thoughts, for example, or of actions. Events happen at specific moments, and they vanish as soon as they appear. In this sense disorganization is a continuous and necessary cause of being. The system has to manage its own *creatio continua*. Thanks to God, Descartes would have said, structures have evolved that make it possible to interconnect events and to limit the choice of the next event.

7. Self-referential systems are paradoxical; their existence implies the unity of different logical levels, of different logical types.[32] To say that their unity produces their unity is not tautological but paradoxical. It is paradoxical because "production" presupposes a

difference between cause and effect, and the autopoietic theory states that the different is the same. The analysis of such a situation requires us to ask how a self-referential system itself handles the paradox of its existence; how it tackles such problems as *the different is the same, the same is different;* how it observes itself as unitary yet multiplex, as a manifold unity; how it simplifies and prepares itself for logical and ontological cognition. This may be seen as an elaboration, by existing, of sufficient identity—or, in the less technical formulation of Paul Valéry, "Je suis né plusieurs, et je suis mort un seul" (Born as several, I die as one).

By now, this may be ununderstandable enough. But whoever gets this message will at least see the possibility of defining the individuality of an individual as autopoiesis. This leads us back to the late scholastic position with which I began. There is no individuality *ab extra*, only self-referential individuality. But this means that cells and societies, maybe physical atoms, certainly immune systems and brains, are all individuals. Conscious systems have no exceptional status. They are a particular type. There is no ultimate, all-encompassing unity. We have a world only in the sense that some autopoietic systems, certainly conscious systems, can conceive of the identity of the difference between themselves and their environments—that the difference is always *one* difference (in distinction to others). Of course, this again does not deny interconnections between systems. As interconnections, however, they have no immanent, natural, or cosmological unity. They are only ecological *relations*. There is no ecological *system*.

This, as it stands, may be a successful scientific theory or even a paradigmatic revolution in systems theory. But does it give an adequate account of what we would like to be when we observe and describe ourselves as individuals? Does it acceptably redescribe our social environment as an autopoietic, self-referential, circular societal system? And is this theoretical reformulation emotionally adequate to the present human condition? We may, of course, define emotions as the autopoietic immune system of the autopoietic psychic system; but again: is this emotionally adequate? Don't we want to need at least a special place, if not the highest rank, for ourselves in the ecology of autopoietic systems—something like "the highest and richest" of all structures of self-reflection, as Gotthard Günther referred to the self-awareness of man?[33]

I see no way to answer these questions with a clear yes or no. But at least we can trace some possibilities, or rather, impossibilities, that may influence future attempts to reconstruct individualism. The most important consequence might well be that the theory of autopoietic systems seems to bar all ways back to an anthropological conception of man.[34] It precludes, in other words, humanism. The reason is simple: there is no autopoietic unity of all the autopoietic systems that compose the human being. Certainly mind and brain never will build one closed, circular, self-referential autopoietic system, because thoughts, as elements of the mind, cannot be identified with single neurophysiological events, as elements of the brain. This is not to deny that we are all human. But to want to be human has no scientific basis. It amounts to sheer dilettantism.

This means that we have to invent new conceptual artificialities in order to give an account of what we see when we meet somebody who looks and behaves like a human being. How do we know that he is one? Because we are self-observers and able within our own self-reference to assume that the other is a self-observer too?[35] Or —the explanation I prefer—because we interpenetrate in a social system that presupposes the other ego?[36]

Moreover, two consequences emerge: first, all observations of individuals (and theories are programs for observation) must focus on difference, not on unity. Otherwise we would not be able to perceive identity. And second, all observations have to choose system references, self-observation being a special case. This means accepting relativism without exception, without any kind of ontological base, without any kind of a priori. The counterbalancing argument can only be that relativism does not preclude, and even is presupposed by, universalism.

Given these constraints, we are free to choose conscious systems as the system reference most appropriate for what we want to express if we claim to be individuals ourselves. This comes very close to what has been done under the heading "transcendental reduction" (Husserl). We drop, however, the distinction between empirical and transcendental. It contradicts the essential unity of the autopoietic process reproducing thoughts out of thoughts (as elements out of elements). Transcendental theory was, after all, a desperate attempt to avoid circularity. The theory of self-referential systems accepts circularity as a basic necessity.

This insight destroys the formula of the individual as the subject. My guess is that the traditional experience (and I say intentionally: "experience") of ennui will provide better clues for a theory of the autopoiesis of conscious systems than the concept of subject did. The seventeenth century made a twin discovery: the subject and its boredom. In other words, the subject has to occupy itself with something, be it economic or aesthetic.[37] Motives, then, are to be thought of as filling the inner void, the empty circularity of pure autopoiesis, of the reproduction of the elements of consciousness by elements of consciousness; and boredom corresponds to the thinking of thinking. During the seventeenth century, both the subject and its ennui became socially acceptable self-descriptions.

Only the theory of autopoietic, self-referential systems seems to be able to formulate this latent unity of the subject and its ennui— a theory of the self-despairing subject, a theory of dynamism achieved through self-desperation—and to formulate it in acceptable terms. At the moment, however, we cannot find even the slightest approach in this direction. We cannot walk a beaten path. We can foresee, however, that it will no longer be possible to use the venerable distinctions between reason, will, and feeling.[38] They have to be replaced by the distinction between autopoiesis and structure. The whole body of knowledge about consciousness, meaning, language, and, above all, "internal speech" will have to be reformulated. There is no dual or even pluralistic self, no "I" distinct from "me," no personal identity distinct from social identity. These conceptions are late nineteenth-century inventions, without sufficient foundation in the facts of consciousness. We simply do not live and do not experience ourselves that way. Moreover, these dualistic or pluralistic paradigms are themselves semantic reactions to the facts of a complex society.[39] We can drop as futile all attempts to reintegrate a decomposed self. If consciousness operates at all, it does so as an individual system, using its own unity and its own conscious events to reproduce its own unity and its own conscious events.

This is the reason why an autopoietic system cannot produce its own end. Humans can commit suicide because the conscious system can interfere with the organic system. But the autopoietic system of consciousness cannot think of death as the last autopoietic element. Autopoiesis is the reproduction of elements that take part in the reproduction of elements, and all attempts to think of a last moment will only produce a *re*productive element.[40] We can be sure that all of this presupposes and has reference to individuality in the sense of a closed, circular, self-referential network, in which the elements of the system are produced by the elements of the system. But beware: this not a nice theory, neither a theory of perfection nor even of the perfectibility of the human race. It is not a theory of healthy states. Autopoietic systems reproduce themselves; they continue their reproduction or not. This makes them individuals. And there is nothing more to say.

Endnotes

1. The more recent school of symbolic interactionism tends to fuse Mead's behavioralism with phenomenology. This fusion cannot succeed and amounts to an enormous hybridization of incompatible kinds of theory. Husserl clearly rejects the idea of internal communication as a model of conscious self-reference, and this rejection is not a casual remark but a necessary precondition for the founding of transcendental phenomenology. See his *Logische Untersuchungen*, 5th ed. (Tübingen: 1968), II, 1, parts 1–8. For a critique and a different (semiological, but again not behavioralistic) position, see Jacques Derrida, *La Voix et Le phénomène* (Paris: 1967) (trans. by David B. Allison as *Speech and Phenomena: and Other Essays on Husserl's Theory of Signs* [Evanston, Ill.: 1973]).

2. Such a perspective may even be historically naive. (See Reinhart Koselleck, *Vergangene Zukunft: Zur Semantik geschichtlicher Zeiten* (Frankfurt: 1979), pp. 38ff.

3. For this point and for a confrontation of American optimism and European scepticism with respect to healthy individuals, see Ray Holland, *Self and Social Context* (New York: 1977).

4. See, within the context of a long discussion about previous theories, Francisco Suárez (1548–1617), *Disputationes metaphysicae*, Disp. 6, esp. 6:14: "Modus substantialis, qui simplex est et suo modo indivisiblis, habet etiam suam individuationem ex se, et non ex aliquo principio ex natura rei a se distincto." *Opera omnia* (Paris, 1866; reprint Hildesheim: 1965), 1:185.

5. Indeed, until the late eighteenth century, human individuals were only a special kind of individual thing *(res)*, characterized by their rational substance. And *res* meant simply a constraint on possible combinations of traits.

6. Much literature is available on this theme. See Alban J. Krailsheimer, *Studies in Self-Interest: From Descartes to La Bruyère* (Oxford: 1962); Niklas Luhmann, "Frühneuzeitliche Anthropologie: Theorietechnische Lösungen für ein Evolutionsproblem der Gesellschaft," in *Gesellschaftsstruktur und Semantik* (Frankfurt: 1980), 1:162–234.

7. Even for nonbelievers, so runs the argument, it would be too risky to ignore this distinction completely.

8. See Kenneth E. Kirk, *The Vision of God: The Christian Doctrine of the Summum Bonum* (London: 1931).

9. Pierre de Villiers, *Pensées et reflexions sur les égaremens des hommes dans la voye du salut*, 3d ed. (Paris: 1700), 2:93.

10. "Qui voudra être devot pour en faire profession, ne le sera pas. Qui le sera véritablement, n'en fera profession sans penser de le faire" *(ibid.,* p. 98). But then, what happens if he knows of, or even hopes for, this inadvertent communication?

11. Many reflections on this point can be found in Pierre Nicole, *Essais de morale* (Paris, 1671–1674; new ed. in 4 vols., 1682). Jesuits, on the other hand, claim to have the special professional skills necessary for making this difficult distinction.

12. See Niklas Luhmann, *Liebe als Passion: Zur Codierung von Intimität* (Frankfurt: 1982), passim, esp. pp. 112ff., 131ff. (English translation: *Love as Passion*, Cambridge: 1986).

13. So the recommendation of the Comte de Versac in Claude Crébillon (fils), *Les Égarements du coeur et de l'esprit* (Paris: 1961).

14. See Francois-Augustin Paradis de Moncrife, *Essais sur la nécessité et les moyens de plaire* (Amsterdam: 1738), pp. 92ff., and many other authors.

15. "Il avoit un quart de loge à l'Opéra, jouoit au lotto et soupoit en Ville." Sénac de Meilhan, *Considérations sur l'esprit et les mœurs* (London: 1787), p. 317.

16. *Ibid.*, p. 41.

17. Perhaps the best theoretical conception can be found in Simon-Nicolas-Henri Linguet, *Theéorie des loix civiles, ou Principes fondamentaux de la société*, 2 vols. (London: 1767). Better known, but less revealing, is the line taken by Voltaire, Rousseau, Diderot, Mercier, and others, who use the distinction between nature and civilization as a springboard for purely moralistic discourses about modern conditions.

18. "Son existence cessa, pour ainsi dire, de lui appartenir." But the institution of property manages to deceive individuals, to "les asservir, sans les empécher de se croire libre" (*ibid.*, 1:198–199). The more optimistic line, and the prevailing one, does not pay any attention to this difference but simply focuses on the institutional unity of property, which gives to all human beings the chance to extend their enjoyment of life. "Die Vergrösserung und Vervielfältigung des Menschenlebens ist der Zweck des moralischen Menschen" [the growth and variation of human life is the aim of the moral man] and individual property is the means necessary to this end. Justice, then, becomes "die unveränderliche Neigung, einem jeden Menschen sein ganzes Eigentum uneingeschränkt zu lassen und unverletzt zu erhalten" [the unchangeable inclination to leave in the unrestricted possession of every man his whole property and to maintain it inviolate]. Johann August Schlettwein, *Grundfeste der Staaten oder Die politische Ökonomie* (Giessen: 1779), pp. 384–385. (Schlettwein was the premier German physiocrat.)

19. Again: descriptions of this process make no reference to religion. Genius is no longer a gift of God and, of course, not a way to salvation but simply an accident of nature, an "arrangement heureux des organes du cerveau." Jean-Baptiste Dubos, *Réflexions critique sur la poesie et sur la peinture*, new ed. (Paris: 1733), 2:7. See also Lodovico A. Muratori, *Della perfetta poesia italiana* (1706; new ed. Milan: 1971), 1:217.

20. "Le coeur s'agite de lui-même et par un mouvement qui précède toute déliberation quand l'objet qu'on lui présente est réellement un objet touchant." Dubos, *Réflexions*, 2:326.

21. See R. G. Saissellin, *Taste in Eighteenth Century France: Critical Reflections on the Origins of Aesthetics, or, An Apology for Amateurs* (Syracuse: 1965).

22. See Alfred Baeumler, *Das Irrationalitätsproblem in der Aesthetik und Logik des 18. Jahrhunderts bis zur Kritik der Urteilskraft*, 2d ed. (Tübingen: 1967).

23. "Etouffer la voix de la nature," says Linguet in *Théorie*, 1:184.

24. To give as much content as possible to humanity within each individual, connecting his ego and the world by the most general, vivid, and free reciprocal action—this is Humboldt's idea of *Bildung*. See "Theorie der Bildung des Menschen," in Wilhelm von Humboldt, *Werke*, 2d ed. (Darmstadt: 1969), 1:234–240, 235n.

25. See K. W. Swart, " 'Individualism' in the Mid-Nineteenth Century (1826–1860)," *Journal of the History of Ideas* (1962), 23:77–90; Stephen Lukes, *Individualism* (Oxford: 1973).

26. See Niklas Luhmann, *The Differentiation of Society* (New York: 1982).

27. For further materials supporting the same point see Luhmann, *Gesellschaftsstruktur und Semantik*, 3 vols. (Frankfurt: 1980, 1981, 1989).

28. Nicolaus Copernicus, *De revolutionibus orbium caelestium libri sex* (1543), ed. Franz Felles and Karl Felles (Munich: 1949), book 1, ch. 10, p. 25, for the system of fixed stars.

29. See Alfred North Whitehead, *Process and Reality: An Essay in Cosmology* (New York: 1929).

30. See Humberto R. Maturana and Francisco J. Varela, *Autopoiesis and Cognition: The Realization of the Living* (Dordrecht: 1980); Francisco J. Varela, *Principles of Biological Autonomy* (New York: 1979); Milan Zeleny, ed., *Autopoiesis, Dissipative Structures and Spontaneous Social Orders* (Boulder, Colo.: 1980); and Milan Zeleny, ed., *Autopoiesis: A Theory of Living Organization* (New York: 1981).

31. See George Spencer Brown, *Laws of Form*, 2d. ed. (London: 1971).

32. See Yves Barel, *Le paradoxe et le système: Essai sur le fantastique social* (Grenoble: 1979).

33. See Gotthard Günther, "Cybernetic Ontology and Transjunctional Operations," in Gotthard Günther, *Beiträge zur Grundlegung einer operationsfähigen Dialektik* (Hamburg: 1976), 1:313–392, 316.

34. But see Edgar Morin, *La méthode*, 2 vols. (Paris: 1977–1980), for the contrary opinion.

35. This is Heinz von Foerster's explanation. See his "Kybernetik einer Erkenntnistheorie," in Wolf D. Keidel, Wolfgang Handler, and Manfred Spreng, eds., *Kybernetik und Bionik* (Munich: 1974), pp. 27–46. See also his "On Constructing Reality," in Wolfgang F. E. Preiser, ed., *Environmental Design Research* (Stroudsburg, Pa.: 1973), 2:35–46.

36. See also Ranulph Glanville, "The Form of Cybernetics: Whitening the Black Box," in *General Systems Research: A Science, a Methodology, a Technology* (Louisville: 1979), pp. 35–42.

37. It is, however, characteristic of the bourgeois drift, which had just begun, that only aesthetic interests were based directly upon the problem of coping with ennui, whereas economic motives were based upon such more "natural" and acceptable sources of motives as uneasiness, restlessness, and unlimited drives.

38. Then, of course, sociologists would have to admit that there is no reference for "voluntaristic action theory."

39. See Jan Hendrik van den Berg, *Divided Existence and Complex Society* (Pittsburgh: 1974). The author, however, would not accept my idea of a semantic reaction.

40. Think of the last-minute reports so fashionable in recent years. But think also of the famous analysis of Jean-Paul Sartre, *L'Être et le néant*, 30th ed. (Paris: 1950), pp. 615–638. Or simply of Paul Valéry's "La mort est une surprise que fait l'inconcevable au concevable." "Rhumbs," in Paul Valéry, Œuvres, ed. Jean Hytier (Paris: 1960), 2:611.

7.
Tautology and Paradox in the Self-Descriptions of Modern Society

Self-referential systems are able to observe themselves. By using a fundamental distinction schema to delineate their self-identities, they can direct their own operations toward their self-identities. This may occur for different reasons and involve very different distinctions. As soon as the need arises to direct self-observations through structural predispositions instead of entirely leaving them to particular situations, we may speak of "self-descriptions." Descriptions fix a structure or a "text" for possible observations which can now be made more systematically, remembered and handed down more easily, and which can now be connected better to each other. Independent and occasional self-observations are not excluded thereby but become less important. Occasional observations now form a "variety pool" for the selection of self-descriptions that can be tested during the evolution of ideas and may be stabilized as tradition. As a result, societies might adhere to traditions of self-descriptions that have lost their adequacy with respect to the structural complexity of the system but that cannot be abandoned since self-descriptions perform important systemic functions.

Thus, in relationship to systemic environments, social-structural and semantic components of a system are not necessarily synchronous. But by and large it seems safe to assume that obsolescence of self-descriptions and misdirection of self-observations will finally become apparent, that a considerable degree of discrepancy cannot be tolerated for a long time, and that a loss of realism in self-descriptions gives reasons for revisions even if the original level of plausibility of the cultural tradition cannot be regained very rapidly. In any case, if one intends to observe and describe how societal self-descriptions and self-observations are transformed in response to structural transformations of society, a broad perspective and a correspondingly abstract theory appear appropriate given such transition periods.

In simple segmentary societies, self-descriptions were rather unproblematic. The level of semantic complexity could be kept fairly low since these societies were organized around very small units such as households, tribes, and settlements and since more complex associations had to function only occasionally. Elementary knowledge of the surrounding geographical space, of individual persons, and—sometimes—of mythologies demarcating the given order of human life from frightening alternative orders were sufficient. Myths and cult forms could be brought into harmony with environmental conditions, structures, and interests without this process becoming visible as a contingent decision. For example, as John Middleton and David Trait state: "Whereas the ancestral cult in particular is a ritualization of organization based on descent, the earth cult is a ritualization of organization based primarily on locality or community with a high degree of political interdependence of descent groups."[1]

Semantic complexity does not increase until society is based more on asymmetries and inequalities. The improbability of the social order becomes apparent and requires explication, if not justification, as soon as center and periphery, particularly urban and rural areas, are separated. This is especially true in the case of hierarchical stratification. Viewed in retrospect, it may seem as if these inequalities exerted some pressures for legitimation of privileged status positions, but this was hardly the case. If social structure differentiates along these lines, legitimation is unnecessary since an alternative order cannot be realistically imagined anyway. Consequently, one can hardly assume "consensus" or the "need for consensus"; as if social order were based on a conscious selection from other possibilities. Articulating the meaning and the "good forms" of social life was purely a matter of the upper classes, i.e., an urban phenomenon. Societal self-descriptions were phrased in terms of *polis-civitas-civilitas-societas civilis,* in religious terms of the corpus Christi or of the "community of sinners" with different prospects for salvation, or in terms of corporate doctrine *(Stände-*

lehre) with its codified morality. But all of these self-descriptions utilized the asymmetric structure of society itself, regardless of whether they emerged from the center or were imposed as self-conceptions of the upper classes.[2]

The most conspicuous characteristic of this pattern of relating social structure to the self-description of society is the opportunity of an unchallenged representation of society in society. There is only one position from which to develop and circulate self-descriptions: the position of the center or of the hierarchical leaders, i.e., the position of the city or of the aristocracy. The asymmetrical form of social differentiation credibly and effectively excludes other possibilities. Under these circumstances, the differences between a primarily religious and a primarily political concept of society cannot be balanced. They are adopted into the cultural semantics and are often structured hierarchically themselves: while gaining priority in the cultural semantics, religion must actually connect to the political center in order to be generally accepted. Viewed retrospectively, this division of labor was important to contemporary experience, but it does not distinguish high cultures of this type from today's society. Rather, the crucial historical difference between past and present society is that the possibility of an unchallenged representation of society in society had to be abandoned upon transition to a primarily functional mode of social differentiation. None of the functional systems can now claim a privileged position; each develops its own description of society according to the presumed priority of its own function. But since the concrete operations of particular systems are too diverse, no system can impose its descriptions upon others. Even if a new type of difference develops, i.e., the difference between functionally differentiated systems and the protest against functional differentiation or, to speak with Habermas,[3] the difference between systems and the life-world, it is impossible to decide from which of the two perspectives society could be described comprehensively or, at least, representatively.

In historical comparison, a characteristic feature of modern society is thus the loss of natural representation or, to use an older term, the impossibility of a *representatio identitatis*. The totality of society is never fully present and cannot be realized as a totality. As a consequence, the concept of representation is reconstructed as a specifically political concept, implying that from now on representation can only be organized according to the functionally limited view of the political system. The question arises of how a self-description of society is possible at all once natural representation must be abandoned.

The underlying hypothesis of the following argument is that society responds to the loss of its natural and unchallenged representation by stating the problem of identity in a more abstract way. It is well known that in the eighteenth century the apotheosis of Reason was suggested as a solution. But this attempt to describe modern society comprehensively has failed. Identifying society in terms of absolute Reason has remained entirely ineffective and generates counterintuitive effects upon implementation into social reality. Restating the problem of representation in a more abstract way, however, corresponds to the nature of the task and—in classical sociological terms—to the necessity to respond to increased differentiation by generalizing the cultural symbolisms expressing societal unity.[4] But formalization and "proceduralization" of the principles of Reason are only expedients that do not promise concrete results once specifications are required. Eventually, only the options of counterfactually adhering to reason, of mere stubbornness, of lament, or of resignation remain. Of course, it is hard to renounce Reason. But perhaps we just remain loyal to an historical name brand of cultural semantics while reality has changed in the meantime. In any case, it should be worth looking for alternatives, for functional equivalents of a Reason-oriented reflection of societal unity.

The apotheosis of the "Ego" in German idealism, particularly following Fichte's doctrine of science, has also failed to provide an adequate societal self-description. The "Ego," however, was already very accurately conceived of as the resolution of a paradox by means of an approximative idea: the Ego posits the difference between Ego and non-Ego and raises itself to an ideal Ego "above and within the limit."[5] But even more so for the orientation toward Reason, the social dimension was lost in the process. The problem of paradox was related to knowledge, not to society. Consequently, the elaboration of theory was preoccupied with religious or aesthetic and, eventually, with pedagogic issues but not with issues of economics and politics.

First, we want to surpass the formalism of Reason and the idealism of the Ego by means of a radical reflection. There can be two different forms of reflecting upon the identity of a system: tautological and paradoxical forms. Correspondingly, we might say that society is what it is or, alternatively, society is what it is not. Both forms of reflection, however, do not improve but block the observations of the system. As with Reason formerly, both forms of reflection lack concrete conceptual and normative implications for possible societal self-descriptions. Since both versions have the disadvantage of sterility, an observer can neither guess which one will be chosen nor recommend which one to choose; nor can he predict what consequences one or the other version of self-description will have for the system.[6] Therefore, even observations of observations or descriptions of descriptions of a system contribute to systemic self-blocking and become tautological or paradoxical themselves since they conceptualize their subject matter in a way that excludes concrete elaboration of societal self-descriptions.

We may overcome this obstacle by examining how the system itself manages to overcome it. In a very general sense, systems avoid tautological or paradoxical obstacles to meaningful self-descriptions by "unfolding" self-reference.[7] That is, the (positive or negative) circularity of self-reference is interrupted and interpreted in a way that cannot—in the last analysis—be accounted for. The most famous example is the type-theoretical solution to the paradox in set theory. In any case, processes of "detautologization" and "de-paradoxization" require the "invisibility" of the underlying systemic functions and problems.[8] That is, nontautological and nonparadoxical societal self-descriptions are not due to individual plans or intentions but are possible only if crucial systemic processes and operations remain latent. Only an observer is able to realize what systems themselves are unable to realize. Or, alternatively, we can say that the problem is to avoid "strange loops," "tangled hierarchies,"[9] or their effects, such as the "double bind", without being able to eliminate tautologies and paradoxes as identity problems of self-referential systems.

Therefore, modern society does not admit that its self-description faces a problem of tautology or paradox. Only by coding its identity is society able to construct social theories. But depending on whether tautological or paradoxical approaches to self-descriptions are selected, very different semantic systems emerge. Tautological approaches lead to rather conservative self-descriptions; approaches based on paradox lead to rather progressive—if not revolutionary—self-descriptions. The basic problem of self-reference generates the antagonism between the two approaches. If society is supposed to be what it is, then the problem can only be to conserve society, to continue solving its problems, and possibly to improve problem solving and to overcome unexpected difficulties. If, on the other hand, society is assumed to be what it is not, then theories of a different kind must be suggested. For example, as popular versions of Marxism or the Cargo cult show, it is possible to define societal identity as a future possibility the realization of which is prevented by certain forces. Alternatively, the problem is restated in a temporally asymmetric way. One then assumes that a structural-logical development will realize—through revolution or evolution— what present society is "not yet."

Elaborated as comprehensive theory, each version faces specific difficulties that need not further concern us here. I am not so much interested in the differences between these two versions or in the amount of intellectual sophistication invested in their conceptual elaboration as I am in the characteristics they share despite, or even because of, the bifurcation between them. What tautological and paradoxical approaches to societal self-descriptions have in common is that they transform descriptions of society into ideologies.

After originating around 1800, the concept of ideology has undergone multiple transformations.[10] First, ideology was thought of as semantic control of social reproduction through ideas. After having been used in a purely pejorative and polemic way, the concept of ideology was finally granted social scientific acceptability. This was largely the merit of Marx and Engels, not so much because they created a perfectly adequate concept of ideology, but because of their theory of capitalist society in which the concept is functionally located. Ever since then, the concept of ideology has displayed a particular reflexivity that appears immune to empirical evidence and criticism. Once their latent function is being revealed, ideologies draw upon some kind of "support" that prevents them from decomposing. That is, an ideology is simply propagated as being "biased" ("parteilich") or as practical knowledge, i.e., as theory that has become "praxis." In directing and justifying social action, ideologies become replaceable once different lines of action seem more appropriate, but they can never be destroyed by criticism.

Practical relevance is part of the very nature and self-explanation of ideology. Observing ideological self-descriptions, however, reveals more complicated frames of reference. Ideologies are firmly based on the implicitness of their basic problem definitions, on concealing their intentions, on the latency of their fundamental assumptions. Descriptions of societal self-descriptions face the antagonism of ideologies instead of reflecting upon the more fundamental problems of tautology and paradox. Each ideology may claim to represent a comprehensive holistic system if it is able to explain why competing ideologies exist. With Marx offering the most ambitious model, successful ideologies make it seem unnecessary to recur to the basic (tautological or paradoxical) forms in which societal self-descriptions conceptualize the problem of identity. An ideology stabilizes itself by including its counterideology into its system, and it is hardly more than a variant of this strategy if the conservatives acquiesce in counterenlightenment while indicating their own standpoint through nothing but aperçus to avoid exposure to criticism.

The crucial themes of early ideologies depended on historical accidents such as the French Revolution or on contemporary social problems, particularly on the consequences of rapid industrialization. This new form of reflecting upon societal identity acquired topics and themes from wherever they were available, and this led to historical relativity and to the gradual obsolescence of many opinions regardless of whether socialist or liberalist ideologies were involved. It was nevertheless premature to diagnose the "end of ideologies." Rather, intellectual scepticism and the readiness for trivial moralizations or, if one considers France, the retreat into literary arcanistics have become even more prominent. But fundamentally different forms of societal self-descriptions have not been established. The senility of formerly predominant ideologies is troublesome for its respective adherents but does not necessarily lead to new suggestions. In the long run, it may be possible that second-order systems-theoretical observations (that is, observations and descriptions of societal self-observations and self-descriptions) will yield very different results. At present, however, there is no elaborated semantic system of this kind that was already implemented as actual societal knowledge. For now, we can only attempt to clarify the implications of this view and to reformulate the concept of ideology from this perspective.

In the nineteenth century, various fundamental distinction schemas had been developed as structural frameworks for particular ideological contents. The self-identification of society requires descriptions based upon fundamental distinctions that define what society is by determining what it is not. Following the distinction between power and property that had already been prominent in the eighteenth century, one suggestion pointed to the distinction between state and society. After rejecting Hegel's attempt to overcome this distinction and to ingeniously suppress the necessarily reemerging problem of paradox, the distinction between state and society was accepted as indisputable fact around the mid-nineteenth century.[11] Depending on ideological predispositions, this distinction made it possible to ascribe more or fewer responsibilities to the state. As opposed to medieval thought, however, the distinction between state and society also made it possible to replace religion with the economy and to describe the future prospects of the epoch in terms of the economy (which was referred to as "society" for this very reason). In this way the concept of society was established, but the (tautological or paradoxical) position pointing at the identity of the difference between state and society was left unoccupied. The old European *societas civilis*—which was still remembered in Hegel's concept of the state—did not find any successors.[12] The fascinating capacities of distinction schemas to direct information processing block insight into the fundamental unity of what is being distinguished: distinguishing what is different makes sense only when positing an underlying identity that permits realizing what is different.

Although the distinction between state and society defined the concept of society and made its usage almost impossible for emerging sociology, the "true" theory of society was built upon a distinction that transcended society as a social system. This distinction related the social—in various terms—to the individual and thus conceived of the individual as an extrasocial entity.

This line of theorizing is already present in the newer philosophy of mind and its notion of the subject. It is also present in a notion of the individual that abstracts from all of its social positions and involvements while presupposing that the individual identifies itself only by reflecting upon its individuality.[13] Such an individual can be expected to live in plural contexts.[14] To compensate for this abstract status, such an individual is capable of complaining—about alienation or about unrealized promises of freedom, about inequality or about the inability of society to live up to the standards that the individual believes all individuals accept as reasonable. All of this shows that the individual is no longer understood as a unique part of nature but in opposition to society.

Conversely, society can then be distinguished from the individual as a collectivity.[15] "Collectivity" may refer to very different ideas: to the population of human beings, to nations, to social order, or to historically varying social formations such as "capitalism." Conceptual clarity, however, is less important here than the underlying distinction that permits identifying the "nature of the social"—the distinction between society or collectivity and individuality. Once this basic distinction is accepted, information can be processed, and various ideological self-descriptions of society can express the concrete meaning and implications of this distinction. That is, since society and individuality are identified in opposition to each other, individual dissatisfaction with society can be expressed in diverse ideological contexts (e.g., conservative or progressive contexts). The basic distinction between society and individuality leaves sufficient room for ideological systematizations of dissatisfaction and hopelessness without excluding the possibility of changes. In the second half of the nineteenth century, this distinction finally ruins the belief in natural progress and inspires the newly emerging sociology to analyze the individual within the structural contexts of modern society.[16]

Comparing the two distinctions between state/society and individual/collectivity, it is easy to see that both fulfill the same function: making information processing possible while at the same time blocking insight into the unity of the differences posited and, thus, into the problems of tautology and paradox. The processing of information in the ongoing self-observation of society utilizes, confirms, and transforms these "distinction directrices." What can be distinguished by means of these distinctions will become "information." But while the distinction between state and society does not transcend the social order and can thus only identify subsystems, the distinction between the individual and society posits an eternal reference point that serves as a standard for evaluating social conditions. Up to the era of the "Frankfurt school," this procedure created a theory of society that was based on the opposition between individual and society. This basic framework today is abandoned and replaced by new distinctions such as between work and interaction or between system and life-world and the "intersubjectivation" and "proceduralization" of the subject (Habermas). But these new distinctions are only different ways to conceal the crucial unity underlying all distinctions.[17]

But let us return once more to the problem of ideology. In his programmatic introduction to the dictionary *Geschichtliche Grundbegriffe*, Koselleck assumes that since the mid-eighteenth century, old European social and political semantics have undergone a fundamental change of meaning.[18] Koselleck argues that the characteristic feature of this change in meaning was the fact that many concepts were now understood as being temporal and ideological.

Koselleck's argument supports the stronger hypothesis that ideologies are in fact characterized by temporalization. Ideologies replace the reference to nature by the reference to historical time and to the present condition of society.[19] In a certain sense, temporalization and "ideologization" cooperate in compensating for the loss of realism that becomes unavoidable as soon as social structure is no longer compatible with only one privileged representation.

Differences in references to historical time reflect differences between conservative and progressive ideologies. To be sure, the experience of accelerating social change sabotages the simple antogonism between progressive and conservative ideologies and transforms it into the question of whether the basic internal dynamics of society such as market economy or uncensored scientific research should be maintained or whether some control is necessary here to preserve elementary human interests. Rapid social change may lead to an exchange of topics between the left and right wings so that formerly conservative topics such as cultural pessimism, critique of technology, and resort to the "state" are now primarily discussed in the leftist camp.[20] But reference to historical time and interpretations of the present social condition are still crucial for distinguishing ideological options (since they must not, as you recall, refer to differences between tautological and paradoxical identity reflections). The exchange of topics proves that concrete ideological commitments are not that important; they only serve to implement a more fundamental antagonism that must not reveal its rationale and that is therefore constantly actualized in the form of interpretations of the present situation of society.

The temporalization of societal self-descriptions and the observation of rapid social change attack the distinction between progressive and conservative orientations. The conservatives start out with disappointment, the progressives end up with disappointment, and both suffer from time and agree therein. The crisis is now ubiquitous. In the extreme case the self-description of society boils down to a "definition of the situation" which always leaves room for controversy, even if the data are indisputable. Depending on expectations and respective ideological opponents, a given level of social welfare is either fairly remarkable or insufficient. Therefore, one either finds reasons to point to the costs and unintended consequences of additional expenditures or to stir up the demand, and a controversy results even if the facts are agreed upon. Similarly, as the debate on "postmodernity" illustrates, intellectual reflection degenerates. The progressive side laments about their goals not being taken seriously anymore and switches from "not yet" to "not anymore," while the conservative side profits from this change and is therefore in the position to renounce further reflection. Such an extensive temporalization still achieves what we expect of ideologies: to deparadoxize or to detautologize societal identity. That is, as opposed to "pure" tautologies and paradoxes, ideologies offer specific descriptions of society and recommend particular programs for action. But the debate on postmodernity will soon become boring since it does not explore new lines of thought and since it simply lets the present time pass.

Once comprehensive self-descriptions of society become more problematic because of the transition to functional differentiation, changes can not only be observed in the temporal but also in the factual dimension. As a result of the loss of a natural and unchallenged representation, society has to deal with a larger amount of contingency. That is, although no societal macroactor can safely be identified, "decisions" become more important than ever before. Particularly, market order and democracy institutionalize more selective decisions. Correspondingly, paradoxes are treated as *moral* paradoxes, implying that they are observed as paradoxes resulting from decisions. The market economy demonstrates that morally reprehensible, egoistic, profit-oriented behavior may nevertheless have virtuous consequences. The opposite is true for politics directed at public opinion. The French Revolution tells its "conservative" observers that the best intentions may have the worst consequences. Thus, in its moral version, the paradox is reversely distributed to the economy and to politics, to society and the state, and is provided there with corresponding institutions (e.g., contractual liberties, elections). The program for Restoration (which is paradoxical itself) is institutionalization of freedom.

But if institutionalization of freedom is adopted as the program for political decisions and for ideology (implying, at first, societal control by means of ideas), an unprecedented need for new semantic certainties emerges. There must be an "inviolate level" of order[21] that resists the play of contingencies and that is not disturbed but reaffirmed by paradoxes and tautologies. There is no position outside of society from which to communicate, but a system can internally test semantic references which may be treated as absolute. This is the starting point for the semantic career of the concept of "values" around the mid-nineteenth century.

Values are "blind spots" that make it possible for systems to observe and act.[22] The value-ladenness of a value defines the position from which to observe, demand, formulate interests, and prepare to act. Distinctions between values and alternative values or between values and undesirable conditions are required for observing. Similarly, actions require that values be included into the semantic systems stabilizing motivations. Values improve the depth, accuracy, and range of observation and orientation but at the same time invite others to observe one's observations and orientations. Values do not express consensus but motivate others to observe one's observations critically.

The conceptual history of "value" has not yet been thoroughly analyzed. However, the concept is not very likely to directly stem from the aristocratic ethos of "valeur."[23] Rather, it is more likely to originate in economics since economics has always attempted to ground the contingency and flexibility of prices in a more stable sphere of underlying values.[24] Basically, the only problem was to generalize this functional context, and since the mid-nineteenth century, this generalization has occurred through extending the concept of value to moral, literary, and aesthetic areas.[25] Eventually, the concept of value denotes preferences the validity of which can safely be assumed in social communication without having to face disagreement. As proven "eigenvalues" of the system, values turn out to be stable even in the context of self-referential operations.[26]

Apparently, the most conspicuous characteristic of values is that they can be communicated inconspicuously. Corresponding to their presumed absolute validity, values are implied as allusions in social communication. One does not tell others that one favors justice; one simply demands more justice in the distribution of income. While communication itself focuses on issues that can be negotiated and disagreed upon, values remain latent in communication. Values are reproduced and stabilized through indirect communication. Since sufficiently general values are easy to find, the latency of values can also be used tactically to suppress disagreement.

One cannot question the validity of values, but one can interpret them. Modern hermeneutics seems to have been invented as a pendant to the new sphere of inviolate values. Emerging as some kind of reflexive theory of the religious system, then subjectified, and finally turning into a philology, modern hermeneutics channels the indisputable into the form of a circle ("hermeneutic circle") in which it is able to orient itself. The hermeneutic approach also denotes an elaborate form of tautology and paradox or an unfolding of self-reference. More specifically, the hermeneutic form of self-reference is chosen if one wants to avoid the option for one or another ideological discourse.

However, the hermeneutic approach to the concept of value has its own consequences and costs. Contrary to common belief, hermeneutics deprives the concept of value of its practical significance. It symbolizes the autopoiesis of communication—but nothing more. The hermeneutic approach does not permit inferring correct behavior, since this would require a resolution of value conflicts that always remains contingent and cannot itself be safely grounded in an "inviolate level" of values.[27] This latter argument is only an alternative formulation for the generally accepted insight that there is no transitive order of values that could be validated as a preestablished hierarchy; that is, regardless of particular circumstances.

In the area of values, the reduction of time horizons to a "definition of the situation" corresponds to what has been observed—partly in very misleading terms such as "postmaterialist"—as global value change. Apparently, this value change is based on a rapidly growing awareness of global risks that is being nourished by the ecological problems of modern society as well as by the difficulties in maintaining the level of social welfare.[28] Particularly in the form of fears or concerns for others or for everyone, anxiety is no longer taboo but a public issue: one could even characterize present times as the "era of unmasked anxiety."[29] This characterization does not imply assumptions about the states of mind of concrete individuals, but it does tell us something about value references in public rhetoric. As a public issue, anxiety advances to a substitute a priori. That is, anxiety cannot be disputed, refuted, or cured. It always appears authentic in communication. It is impossible to reply "you are wrong" to someone saying he is afraid. Anxiety thus deserves and creates respect or at least tolerance; it makes disagreement incommunicable and thus serves as the focus for the "new values."

At the same time, anxiety blocks insight into the problems of tautology and paradox, the pure reflection of which, as stated before, would block communication. Anxiety, on the other hand, releases communication, with its new values profiting from the relief this release provides. Anxiety releasing communication may even lead to a previously unknown form of unreasonable loquacity. But still, we can detect here a way of deparadoxizing societal identity problems that no longer requires an ideology in the classical sense of the term. Ideologies have always been required to offer more than simple value recommendations. They were equipped with cognitive components, i.e., with descriptions of social conditions and problems. Possibly, the cognitive component can now be reduced to the universal formula of anxiety directing the selection of descriptions, "scenarios," world models, and general summons. But this formula would terminate the self-description of society before it detected its own arbitrariness.

Until now I have dealt with tautologies and paradoxes as logically equivalent yet reversed schemas of observations and descriptions. However, this assumption turns out to be problematic if one regards tautologies as special cases of paradoxes. Indeed, tautologies turn out to be paradoxes, while the reverse is not true.

Tautologies are distinctions that do not distinguish. They explicitly negate that what they distinguish really makes a difference. Tautologies thus block observations. They are always based on a dual observation schema: something is what it is. This statement, however, negates the posited duality and asserts an identity. Tautologies thus negate what makes them possible in the first place, and, therefore, the negation itself becomes meaningless.

If we take this consideration seriously, we can no longer assume a functional equivalence of tautologies and paradoxes or of deparadoxization and detautologization. We can then account for the frequently reported observation that the intellect has a certain preference for the leftist side of the political and intellectual spectrum. Apparently, it is more productive to deal with the resolution of paradoxes than with the unfolding of tautologies (which should not lead to the wrong conclusion that deparadoxization generates true ideological knowledge).

Therefore, there is all the more reason to conduct further research on ideologies, on temporalizations, and possibly on other solutions to the problem of self-descriptions based on paradox and deparadoxization. The main question underlying this research would be under what conditions can deparadoxization be developed productively instead of pathologically or as a creative instead of a vicious circle.[30] Since all self-descriptions of society are either based on paradox or on tautology, the problem is not to avoid paradox or tautology but to interrupt self-referential reflection so as to avoid *pure* tautologies and paradoxes and to suggest meaningful societal self-descriptions. The well-known problem of "harmless" self-references (following the pattern of "this sentence is a sentence") becomes less important then, since, as I stated, these appear paradoxical to an observer, too, and thus must be dealt with in the same manner as paradoxes. That is, detautologizations are deparadoxizations, and in both cases the problem is to transform infinite into finite information charges. Correspondingly, the logical-mathematical way of dealing with this problem will have to be revised: paradoxes do not originate (inevitable) vicious circles, but such circles will result from unsuccessful attempts to deparadoxize.[31]

Each observation of systems that observe themselves face the question of the inherent limits to self-referential operations. Especially in the philosophical theory of truth, it is a well-known fact that allowing for unrestricted self-reference leads to tautologies and paradoxes. In observing and describing the self-descriptions of modern society, one gains the impression that modern society confronts this problem but is unable to realize it *as* a problem.

Several evasive strategies can be observed. One of them, the discourse on the "subject," simulates the problem for a case that is external to society. This makes it possible for society to cultivate the illusion of being constituted in a deficient but not paradoxical way. Even Jürgen Habermas' (1985) brilliant and keen exposure of the paradoxical self-enlightenment of the subject depends on the externalization of self-reference. Habermas presents the paradigm of communicative intersubjectivity as a regulative ideal implied in communication itself—as if only the subjects posting their own rationality faced the problem of self-reference. But since we know that unrestricted self-reference is impossible for purely logical reasons, the idealization of intersubjective communication will only interpret the process of self-referential constitution, and then the question arises: why does self-reference have to operate in this way but not another?[32]

This evaluation of Habermas' theory corresponds to the results of my previous argument. Societal self-descriptions that are unable to describe what is in fact indescribable makes use of semantic expedients that conceal this very fact but nevertheless permit self-description. Certain distinctions that identify society in opposition to something else (the "state," "Gemeinschaft," or the "individual") have performed alibi functions in this way. Ideologization and temporalization make this alibi function precarious but not transparent. Values provide the corresponding explanation: new "inviolate levels" are required if everything appears to be contingent and if communication itself must test what will work as a starting point for the unfolding of self-reference.

The question arises of how "sociological enlightenment" observing and describing the self-referential constitution of society is still possible under these circumstances. Which semantic system can be stabilized in such a process of describing descriptions of descriptions? Especially, what are the implications of the fact that societal observations and descriptions are possible only as self-observations and self-descriptions, since no individual mind could ever be a "subject" in the sense that it was the only basis for such observations and descriptions?

The basis for answering this question lies in the assumption that in society there are no unobserved operations—similar to the observation that communication cannot be terminated by communication. Maintaining and continuing the autopoiesis of society, communication is always observed in terms of distinctions that apply to both communication and observation (for example, *this* has been said and not what I expected). On this factual basis the difference between operations and observations can be assumed to be universal and to perpetually reproduce itself. The autopoiesis of society cannot be continued without simultaneously creating new possibilities for observation.[33] The universal validity of this hypothesis implies that observations themselves are only possible as autopoietic operations or, in the case of social systems, as communications.

The distinction between "natural" and "artificial" restriction of self-reference is based on this distinction between operations and observations.[34] Interruptions of self-reference that the system regards as necessary conditions for the possibility of its operations may be called "natural." On the other hand, "artificial" restrictions of self-reference are those that are perceived as contingent, i.e., as selections from other alternatives. Natural interruptions of self-reference thus block insight into the paradoxical and tautological problems of self-referential identifications. In fact, they make these problems invisible. Artificial interruptions, on the other hand, allow for this insight but postulate that the paradox be resolved.

The distinction natural/artificial (necessary/contingent) always refers to particular systems. Moreover, it is subject to early changes or learning processes. If a system is able to discover new "inviolate levels" that serve to deparadoxize its identity, semantic systems deemed necessary may become contingent. European Enlightenment was an evolutionary process of this kind that failed, however, to develop its own self-reference through the semantic system of subjective Reason. Furthermore, the distinction necessary/contingent helps explain how previously undoubted foundations of societal semantics are suddenly suspected of being contingent once evolution changes the pattern of social differentiation.

Most importantly, however, this distinction helps clarify the relations between observations (self-observations) and operations and, consequently, the relations between society and its own self-reflections and descriptions. An observer can realize that self-referential systems are constituted in a paradoxical way. This insight itself, however, makes observation impossible, since it postulates an autopoietic system whose autopoiesis is blocked. Therefore, the assumption of pure and unrestricted self-reference transfers the paradox to the observation itself. Such an observation would contradict its own intentions. Therefore, realizing the necessity of interruption in processes of self-referential constitution deparadoxizes the object of observation and—at the same time—the observation itself. Independent from all a priori conditions for the possibility of knowledge, this insight unites the observation and its object and thus makes possible societal self-observations and descriptions.

The distinction between natural and artificial restrictions of self-reference is very important, since it permits maintaining the distinction between observations and operation, although both are possible only as systems (i.e., as unfolded or deparadoxized operations.) The distinction natural/artificial can be utilized in such a way that an observation can interpret as artificial and contingent what the system itself assumes to be natural and necessary. For example, an observer may examine how a system creates the impression of its self-determinations being natural, necessary, and lacking functional alternatives. He may then, for example, search for functional equivalents for the notion of God serving to deparadoxize the religious system (see essay 8 in this volume). To use Heinz von Foerster's (1979) formulation,[35] in this way an observer can see that the observed system cannot see that it is unable to see what it cannot see. This insight marks the real epistemic gain second-order cybernetics has to offer. Any different goal of sociological enlightenment would only lead into the well-known self-contradictions.

I do not claim that von Foerster's formula expresses ultimate truth. But I do claim that it defines the starting point for a theory determining which type of societal self-description can be proven as adequate even if revealed as artificial and contingent.

What is real can also be observed. Therefore, in the course of history experiences gained by certain social formations from reflecting upon themselves accumulate. Since modern society began to observe and describe itself in the eighteenth century, it is obviously more capable of doing so now than ever before. In any case, the negative aspects of modernity that have been observed since the very beginning of the bourgeois movement can now be interpreted neither as transitory phenomena nor as unavoidable costs of the progress of civilization. Before today society has not been fully confronted with the consequences of its structural selections. This is especially true for the ecological problems resulting from its own rationality. Therefore, it seems close at hand to push self-observations and self-descriptions up to the point of an obviously paradoxical conclusion: that one wants what one does not want.

Searching for positions from which to describe modern society, one encounters social movements. Very typically, these movements attempt to fight society from within society just as if they were external to society. After time-consuming and consequential but unsuccessful attempts to organize around a particular phenomenon—key word: "capitalism"—the so-called "new social movements" of today develop a much more radical perspective and thus fit into an historical situation that provides better opportunities for self-descriptions. New social movements pursue broader concerns and, therefore, draw upon more heterogeneous motivations (which has thwarted many attempts to interpret them as one unified movement). They are radical and nonradical at the same time. They are concerned with preserving single trees and with societal change, with avoiding nonnatural radioactivity and with a different form of life. Often, these movements display contradictory orientations. For example, they pursue ecological goals while criticizing the purely economic rationality of their opponents. Or these movements are internally cleaved. For example, the issue of equality put forward by the women's movement articulates a purely bourgeois demand, while the search for a semantics of femininity expresses the concern for a wholly different form of life. These movements embody the possibility of a critique of society that is much more radical than anything Marx could envision and dare. They are broadly concerned with many consequences of the differentiation of functional systems, and if they do have a radical intention, then it would be the critique of functional differentiation.

The critique of functional differentiation, however, reaches the limits of alternativity. A society can imagine a change in its principle of stability, in its pattern of differentiation or of drawing systemic boundaries as nothing but catastrophe. Like the past critique of the feudal order, the critique of functional differentiation remains a moral critique that is unable to indicate alternative lines of evolution. It is indisputable that improvements can always be made and that people always sin against other people. But in this way the new social movements become inevitably preoccupied with the issues of the day and are different only in their less complicated ways of dealing with them. Their counterpublicity depends on a vivid exchange with the bourgeois publicity against which it seeks to demarcate itself. New social movements demand public recognition by overexaggerating their morality and by choosing unconventional techniques of self-presentation. But they are recognized anyway, and this always occurs within society, not against it.

The secret of alternative movements is that they cannot offer any alternatives. They have to conceal this fact from others and from themselves, and in this way they contribute to deparadoxization. And apparently, this contribution turns out to be rather productive.

Endnotes

1. John Middleton and David Trait, eds., *Tribes Without Rulers: Studies in African Segmentary Systems* (London: 1958), p. 25.

2. Therefore, Edward Shils (1961) correctly states that the main problem resulting from the separation between center and periphery is the diffusion of culture. Shils, "Center and Periphery," *The Logic of Personal Knowledge: Essays presented to Michael Polanyi* (London: 1961), pp. 117–131.

3. Jürgen Habermas, *Der Philosophische Diskurs der Moderne: Zwölf Vorlesungen.* (Frankfurt: 1985).

4. Talcott Parsons, "Durkheim's Contribution to the Theory of Integration of Social Systems," in Kurt H. Wolff, ed., *Emile Durkheim, 1858–1917* (Columbus: 1960), pp. 118–153.

5. Johann G. Fichte, "Grundlage der Gesamten Wissenschaftslehre," in Johann G. Fichte, *Ausgewählte Werke in Sechs Bänden* (Darmstadt: 1962), 1:509.

6. See Frank R. Stockton, *The Lady, or the Tiger?, and Other Stories* (New York: 1969) for a discussion of such a dilemma arising for an observer.

7. Lars Löfgren, "Some Fundamental Views of General Systems and the Hempel Paradox," *International Journal of General Systems* (1978), vol. 4; Lars

Löfgren, "Unfoldment of Self-Reference in Logic and Computer Science," *Proceedings of the Fifth Scandinavian Logic Symposium* (Aalborg: 1979).

8. Yves Barel, "De la ferméture à l'ouverture en passant par l'autonomie?" in Paul Dumouchel and Jean-Pierre Dupuy, eds., *L'Auto-Organisation: De la physique au politique* (Paris: 1983), pp. 466–475.

9. Douglas R. Hofstadter, *Gödel, Escher, Bach: An Eternal Golden Braid* (Hassock, Sussex, England: 1979).

10. Ulrich Dierse, "Ideologie," in Joachim Ritter, ed., *Historisches Wörterbuch der Philosophie* (Basel and Stuttgart: 1976), 4:181–185); and Ulrich Dierse, "Ideologie," in Otto Brunner et al., eds., *Geschichtliche Grundbegriffe: Historiches Lexikon zur Politisch-Sozialen Sprache in Deutschland* (Stuttgart: 1982), 3:131–168.

11. Ernst-Wolfgang Böckenförde, "Lorenz von-Stein als Theoretiker der Bewegung von Staat und Gesellschaft zum Sozialstaat," in *Staat und Gesellschaft* (Darmstadt: 1976), pp. 131–171. Lorenz von Stein, *Geschichte der Sozialen Bewegung in Frankfurt von 1789 bis auf Unsere Tage* (Leipzig: 1850).

12. Niklas Luhmann, *Die Unerscheidung von "Staat und Gesellschaft": Vortrag auf dem 12th Weltkongress für Rechts- und Sozialphilosophie* (Athens: 1985).

13. In passing I want to note that the constitution of individuality also generates typical tautological and paradoxical problems of self-referential identification. Similarly, strategies of detautologization and deparadoxization and of meaningful self-identification have to be devised for personal systems. As an example, consider the semantic career of the "unconscious" defining individuality or theories of dual or multiple, i.e., person *and* social identities.

14. Loredana Sciolla, "Differenziazione Simbolica e Identità," *Rassegna Italiana di Sociologia* (1983), 24:41–77.

15. The ideological opposition of individualism to collectivism did not occur before the second half of the nineteenth century. Apparently, this distinction represents the attempt to differentiate further between the positions opposing individualism (socialism/communism/collectivism); see Rauscher, "Kollektivismus, Kollektiv," in Joachim Ritter, ed., *Historisches Worterbuch der Philosophie* (Basel and Stuttgart: 1979), 4:884f.

16. Otthein Rammstedt, "Zweifel am Fortschritt und Hoffen aufs Individuum," *Soziale Welt* (1985), 36:483–502.

17. See Habermas' stern critique *(Der Philosphische Diskurs)* of the paradoxical consequences of a subject-oriented program for enlightenment. But his critique raises the question of whether there can be a nonparadoxical concept. Habermas claims that the new paradigm of communicative understanding avoids problems of paradox, but how can self-reference be restricted so as to avoid paradoxes? It seems to me that the *historical* analysis of the self-enlightening subject and its opponents prevents a sufficiently *abstract* analysis of the paradoxical problems of self-referential systems in general. The paradigm of communicative understanding is presented in a polemic way that hardly evidences readiness for communication and understanding. As a description of societal self-descriptions, Habermas' paradigm turns paradoxical itself. And it is precisely this problem that is addressed by the mode of observation I call—following Heinz von Foerster—"second-order cybernetics."

18. Reinhart Koselleck, "Einleitung," in Otto Brunner, Werner Conze, and Reinhart Koselleck, eds., *Geschichtliche Grundbegriffe: Historisches Lexikon zur Politisch-Sozialen Sprache in Deutschland* (Stuttgart: 1972), 1:xiii–xvii.

19. Jürgen Wilke, *Das "Zeitgedicht": Seine Herkunft und frühe Ausbildung* (Meisenheim: 1974).

20. See Ortwin Renn, "Die Alternative Bewegung: Eine historisch-soziologische Analyse des Protestes gegen die Industriegesellschaft," *Zeitschrift für Politik* (1985), 32:153–194, regarding the critique of technology; and see Dieter Grimm, "Reformalisierung des Rechtsstaates als Demokratiepostulat," *Juristische Schulung* (1980), 20:704–709, regarding positive law and the state.

21. Hofstadter, *Gödel, Escher, Bach*.

22. William James, "On a Certain Blindness in Human Beings," in *The Works of William James* (Cambridge: 1983), pp. 132–149.

23. Abbé Morellet, *Prospectus d'un Dictionnaire de Commerce* (1769, reprint, Munich: 1980) though, observes such a change from "force" or "vigeur" to "utilité" in the eighteenth century.

24. Of course this attempt was made in very different theories; see Rudolf Kaulla, *Die geschichtliche Entwicklung der modernen Werttheorien* (Tübingen: 1906).

25. A very broad concept of value can already be observed in the eighteenth century (as an example, see Jacques Pernetti, *Les Conseils de l'amitie*, 2d ed. [Frankfurt: 1748], whose definition comprises obligations and pleasures, honor and life, health and wealth), but it was located in the context of a utilitarian anthropology.

26. Heinz von Foerster, *Observing Systems* (Seaside, Calif.: 1981).

27. See Niklas Luhmann, *Soziale Systeme: Grundriβ einer allgemeinen Theorie* (Frankfurt: 1984), for a discussion of the necessity for distinguishing between values and programs.

28. "The main principle of the new values appears to be that of avoiding risks." Walter L. Bühl, *Ökologische Knappheit: Gesellschaftliche und technologische Bedingungen ihrer Bewältigung* (Göttingen: 1981).

29. Werner D. Fröhlich, *Angst: Gefahrensignale und ihre psychologische Bedeutung* (Munich: 1982).

30. Klaus Krippendorff, "Paradox and Information," in Brenda Derwin and Melvin J. Voigt, eds., *Progress in Communication Sciences*, 5:45–71.

31. Charles S. Chihara, *Ontology and the Vicious-Circle Principle* (Ithaca, N.Y.: Cornell University Press, 1973).

32. Even if it were correct, the argument that this idealization is *implied* in communication does not solve the problem since *self-reference is also implied in communication*.

33. It can be shown that communicating systems identify action by means of attributions in order to observe themselves. Luhmann, *Soziale Systeme*.

34. Löfgren, "Unfoldment of Self-Reference," implicitly proposes this distinction without suggesting that it was logically possible to decide whether one is dealing with natural or artificial forms of deparadoxization.

35. Heinz von Foerster, "Cybernetics of Cybernetics," in Klaus Krippendorff, ed., *Communication and Control in Society* (New York: 1979), pp. 5–8.

8.
Society, Meaning, Religion — Based on Self-Reference

> When we are moved to seem religious
> Only to vent wit, Lord deliver us.
> John Donne
> "A Litany," ll. 188–189

Sociological theory in its present Alexandrian phase seems to be preoccupied with the interpretation of its classical authors.[1] Doing sociology of religion means doing empirical research on presumably religious persons or institutions; and it means returning to Emile Durkheim or Max Weber for theoretical inspiration. Religion, then, is supposed to work as an integrative factor on the level of total societies and as a motivational factor on the level of individuals. At both levels it supplies the meaning of meaning, a meaningful "ultimate reality." All symbols and values that operate at this highest level of last resources can be qualified as religion—a civil religion in the sense of Rousseau or Bellah.

We also know the objections. Religions can stimulate debates and fights. They also have disintegrative effects. Their motivational effect may well be questioning religion itself. It may be a social activity, but also a retreat. Statements about the function of religion resemble proverbs. They always need counterproverbs to be operationally useful.

Years ago Clifford Geertz aired the same complaint about dependence upon classical authors with respect to anthropological research.[2] It may have been a mere accident that his lines were written in an essay on the cultural *system* of religion. But if this coincidence happened only by chance, it was still a significant accident. In fact, systems theory, at that time, was hardly able to deliver the goods. Parsons himself had started by presenting his classical authors. He attempted to show that the difference between society and individual, between social and motivational factors, and between Durkheim and Weber does not matter very much; and that it cannot matter very much in the field of sociology where this very difference is the core problem of theory. This preoccupation with an historical problem, with the split paradigm of individual and society, led Parsons to look for a solution by unfolding the framework of the general action system which could assign appropriate places to the personal system, the social system, and other systems as well. He had to pay foreseeable costs. He had to present his generalizations as a purely analytical framework, based on an *analysis* of the *components* of the *concept* of action. Moreover, to compensate for *generalization*, he needed a technique of *respecification*. His decision was to use cross-tabulation, and we all know the consequences.

The verdict on Parsons, accepted today by public opinion, is a verdict based on an impressionistic evaluation of evidence. It is not based on an adequate understanding of the structural constraints of his theory—or for that matter, of any theory. However, I do not want to found the following considerations on a judgment for or against Parsons. Rather, my point is that, in recent years, general systems theory has taken a fascinating turn toward a general theory of self-referential systems, and I want to explore some of its consequences for a theory of society and a functional analysis of religion.[3]

Self-referential systems are not only self-organizing or self-regulating systems. Recent theoretical innovations use the idea of self-reference also at the level of the elements or components of a system.[4] This means that self-referential—or for that matter, autopoietic—systems produce the elements that they interrelate by the elements that they interrelate. They exist as a closed network of the production of elements which reproduces itself as a network by continuing to produce the elements that are needed to continue to produce the elements.[5]

Societies are a special case of self-referential systems. They presuppose a network of communications, previous communications and further communications and also communications that happen elsewhere. Communications are possible only within a system of communication and this system cannot escape the form of recursive circularity. Its basic events, the single units of communication, are units only by reference to other units within the same system.[6] In consequence, only the structure of this system and not its environment can specify the meaning of communications.

Unlike other types of social systems, societies are encompassing systems, including all communications that are conceived as possible within a given context of communication and excluding everything else—even minds, brains, human beings, animals, natural resources, and so forth. Societies, of course, presuppose an environment. They depend upon their environment. Their autonomy cannot be conceived as independence. It is the self-referential circularity itself—not a desired state of being relatively independent from the environment, but an existential necessity. Whatever can happen as a communicative event produces society, entering into the network of reproducing communication by communication. The system expands and shrinks, depending on what it can afford as communication. It cannot communicate with its environment, because communication is always an internal operation.

Communication systems develop a special way to deal with complexity, i.e., introducing a representation of the complexity of the world into the system. I call this representation of complexity "meaning"—avoiding all subjective, psychological, or transcendental connotations of this term.[7] The function of meaning is to provide access to all possible topics of communication. Meaning places all concrete items into a horizon of further possibilities and finally into the world of all possibilities. Whatever shows up as an actual event refers to other possibilities, to other ways of related actions and experiences within the horizon of further possibilities. Each meaningful item reconstructs the world by the difference between the actual and the possible. Security, however, lies only in the actual. It can be increased only by indirection, by passing on to other meanings while retaining the possibility of returning to its present position. Again, a self-referential, recursive structure is needed to combine complexity and security.

This highly successful arrangement of meaning-based communication is the result of an evolutionary development. It has three important consequences which together build up the basic structures of societies:

> 1. The autopoiesis of communication by communication requires *closure*. Meaning, on the other hand, is a completely open structure, excluding nothing, not even the negation of

meanings. As systems of meaning-based communication societies are closed and open systems. They gain their openness by closure. "L'ouvert s'appuie sur le fermé."[8]

2. Communication and meaning are different ways of creating *redundancy*. Communication creates redundancy by conferring information to other systems. Third parties, then, have a choice of whom to ask.[9] Meaning creates redundancy by implying a surplus of further possibilities which nobody will be able to follow up all at once. In view of this redundancy which is continuously reproduced by meaning-based communication every next step has to be a *selection* out of other possibilities. Within the world created by the operations of this system every concrete item appears as *contingent*, as something that could be different.[10] Societies, therefore, operate within a *paradox world*, the paradox being the *necessity of contingency*.[11]

3. Nothing, of course, is paradox per se—not the world, not nature, not even self-referential systems. To call something "paradox" is nothing but a description, and it is appropriate only if one wants to draw conclusions or use other ways of long-chain reasoning. Paradoxes are obstacles only for certain intentions. The paradoxification of being, therefore, is a sociological correlate of an increasing need for descriptions, particularly for self-descriptions of the societal system, and it seems to indicate that such descriptions have to be used within a complex, highly interdependent semantic framework with problems of logical control.

The plenitude and voidness of a paradoxical world is the ultimate reality of religion. The meaning of meaning is both: richness of references and tautological circularity.

Society can exist only as a self-referential system, it can operate and reproduce communications only within a Gödelian world. This general condition makes "religion" (whatever this means) unavoidable. Social life, therefore, has a religious quality—Georg Simmel would say: a "religioid" quality.[12] The paradoxical constitution of self-reference pervades all social life. It is nevertheless a special problem in social life. The question of the ultimate meaning can be raised at any time and at any occasion—but not all the time. If it can be reduced to one question among others, the meaning of the whole becomes a special problem within the whole. Then, society develops *forms* of coping with this problem, of answering this question, *forms that deparadoxize the world*. Then it becomes possible to focus consciousness and communication on these forms and, by this very fact, it becomes possible to risk negation or to look for other forms. Religious forms incorporate, so to speak, paradoxical meanings; they differentiate religion against other fields of life; they involve the risk of refusal; they inaugurate deviant reproduction, i.e., evolution.

Forms convince by implicit self-reference. They propose themselves. They can be "taken for granted in everyday life" because they resist further decomposition. They enforce a "take it or leave it" decision. They reject development. In this sense they have a ritualistic quality.[13] The ritual represents religion because it corks up self-reference; the ghost has to stay in the bottle. But over time and within the context of social evolution ritualistic forms may become maladaptive. They may retain their religious quality and fulfill their religious function *by remaining maladaptive*.[14] They may, however, find functional alternatives in *increasing the ambiguity of forms*.

Ambiguity of forms comes about, if the problem of form is reconstructed as a problem of the *relation between form and context*. The religious (or aesthetic or whatever) meaning of forms, then, depends upon the way in which the form organizes its context, e.g., the temple organizes the surrounding nature by referring to itself.[15] Ambivalence creeps in, if several views are possible, seeing requires a second look, secrets can be unveiled, "alétheia" (truth as the unveiled) becomes a problem. If such a relation between form and context can be questioned and changed, forms can be preserved within a changing context, for contexts can be used to renew forms. Cults may retain their religious meanings by survival and may transfer their function to a different context; or religious contexts may be used to replace one cult by another, e.g., to build a church in the place of a heathen sanctuary. However ambivalent, the paradox of form is the paradox of organizing *context* by *self-reference*. As long as this is the case, forms can seize and retain a religious meaning and may, at the same time, be exposed to deviant reproduction, i.e., evolution.

Translating this into the language of functional analysis, we can say that the fundamental problem of the paradoxical world can be "solved" (i.e., transformed into minor problems) by religion. Plenitude and voidness is the same, meaningful and meaningless life is the same, order and disorder is the same, because the world can be constituted as unity only. But since we cannot accept this last unity as it is, we have to replace it by easier paradoxes: by forms. Forms that retain this functional relation to the ultimate paradox remain thereby religious forms. Forms that can be observed as referring to the ultimate paradox are accordingly observed as religious forms. And forms that can be described as referring to the ultimate paradox are thereby described as religious forms. There is no other way to identify religion, and there is no room for free play in observing and in describing religion. There are, however, many functional equivalents fulfilling its function, and we may find, within one society, many different degrees of sensibility in observing and describing religion. Thus, particularly in modern society, it may become the job of divine detectives to find out what can be observed and described as referring to religion in the paradoxes of art and love, of sovereign power, of making money by making money, or of recognizing the conditions of cognition.

Special forms require special ways to treat them. The ways to encounter them, to avoid them, to behave in their presence are part of their context, therefore part of their meaning. From a structural point of view, the differentiation of forms with specific religious functions inaugurates the development of a special social system serving religious goals. The history of religion is the history of its differentiation.

A theory of religious evolution does not need to be written in terms of a phase model of religious development.[16] It is even questionable whether the theory of evolution can ever arrange history in the form of Guttman scales or any other kinds of linear succession.[17] The theory of evolution tries to explain the possibility of unplanned structural changes; it is not a theory that describes the structure of processes, let alone a theory of a unique process of historical phase-to-phase development. To renounce such an overambitious goal may well be the condition for recombining sociological theory and historical research.

The problem of how to combine a theory of self-referential systems and a neo-Darwinistic theory of evolution is increasingly attracting attention.[18] One possibility might be to conceive of evolution as a *transformation of the paradox of self-reference*. The improbable state of self-referential systems becomes possible and even probable by differentiation—above all by the differentiation of systems and environments. The outcome is the probability of the improbable which, at the same time, is the improbability of the probable.

Translated into a theory of religious evolution, this means that religion becomes endangered by its own success. It is a successful way to handle paradoxes. However, every new form inherits the improbable. It may become normal life, normal society, normal religion; but this does not extinguish the fundamental question of how the unnormal can be normal, of how the improbable can become probable, of how the self-referential circularity can become hierarchy.

Evolution is not a goal-seeking process. Its causes are accidental; they are not appropriate means to produce a result. In other words, the evolution of religious forms and religious systems does not depend on religious causes, events, or experiences (although the religious system will describe its own history in these terms). Since we conceive of society as a self-referential system of communication, we have to suppose that changes in the structure of communication will be one, if not the important change that makes it necessary to adapt religion to new means. The breakthrough may well have been the invention of an easy system of writing, the invention of the alphabet.[19]

By no means does this amount to saying that religion essentially had to be reduced to written communication. The contrary is true. Orality, as a specific way of communication, even gained in importance.[20] The point is that the new facilities of writing and reading did *change the modes and ways in which self-reference is implied in communication*. Referring to another previous or later communication became independent of the spoken word as an actual event.[21] It became independent of the presence of persons, independent of situations, independent of gesticulation and intonation, and above all: independent of the individual and collective memory. It became a matter of arranging the text. Moreover, the written text did preserve everything, important or not, that was written. It was no longer necessary to give special marks to preservable communication, e.g., solemn expression or rhythm. But these had been the traditional ways of religious design. Its form became replaceable. It did not become superfluous. But to the extent that the ways of religious expression were the result of general problems of self-reference in oral communication, this was no longer the case. Solemnity became a matter of linguistic choice and, thereby, a problem of belief.[22]

Therefore, it is not inappropriate to see the elaborate forms of theological semantics and argumentation of later on as the desperate attempt of religion and its professionals to survive in spite of the alphabet. This necessity became a virtue. The theological construction of the Trinity has been invented as the most appropriate reaction: its internal unity achieved by the spoken word which all three components hear at the same time, and its external presentation adapted to the closure of human society as a system of oral and written communication. The technological device itself of writing became sanctified, the gospel was preserved in book form, and this bookish attitude to religion was still reinforced by the invention of printing. The gospel was now accessible to everybody who could read. The Church could no longer present itself as a long chain of oral transmissions; it had to change itself into a system of instructing and supporting reading believers. Again, preaching did not become superfluous; but it had to be good preaching with a view to the fact that all cross-references of the religious belief system were available as written and printed text.

Then, and only then, the ancient ways to formulate religion could be rediscovered as "sublime style," and the eighteenth century pursued this line by inventing the difference between the sublime and the beautiful to make sure that religion (and particularly religious terribleness) could now, as ever, find appropriate forms and would be preserved in spite of aesthetic alternatives.[23]

A further consequence of literacy was even more important. The most immediate result of alphabetic writing was the introduction into evolution of wide discrepancies between semantics and social structure. In a way, the resulting problems were formulated by Plato, but his philosophy itself sided with "ideas."[24] In general, the literature of the Greek city-states became aware of differing realms of meaning, especially of politics and law, knowledge and friendship *(politeîa, nómos, epistéme/dóxa, philía)*;[25] but these differences were no longer representative of social structure; they underrated, for example, economy and religion.

From that time on, and depending upon a technique of easy writing and reading, the increasing probability of the improbable has been generating further complications. For society in general and religion in particular we have to follow two different ways to cope with this dilemma, the one being semantic, the other relying on social structures, e.g., the churches. The discrepancies between these two—the church never becomes a *communio sanctorum*—is one aspect of the problem. It is also the main dynamic factor in religious and perhaps in social history.

It is easy to recognize our problem, if we look at the semantic forms of theological belief that have evolved within the Christian religion. "God" can be seen as the centralized paradox which at the same time deparadoxizes the world. Therefore, we find the asymmetrical notion of creation and, contingent upon this, the idea of the contingency of the world. We have the roots of a hierarchical structure that can be copied everywhere. Original sin symbolizes the beginning of difference and the transformation of the paradox, becoming labor, but remaining difference. The incarnation of God on earth makes the improbable probable. The issue is "salvation," i.e., overcoming difference. But then, salvation again becomes improbable; it becomes contingent upon grace and, finally, in itself turns into an impenetrable and unrecognizable determination. The faith may remain simple, but the belief becomes complex. The theological elaboration uncovers the circular relation between the problem and its solution. It exposes the paradox. It tries to tackle the latter with its own means. And "all was reduc'd to Article and Proposition," as Shaftesbury complains.[26] Whatever we may think of the belief system of this particular religion, it brings about an important structural change—some would say evolutionary advance, or even evolutionary universal[27]—compared with earlier religions. Never before had religion been so articulate. Never before had it set up its own distinction between believers and nonbelievers, abstracting from all other distinctions such as our people/other people, citizens/strangers, or freemen/slaves. Never before was it so completely on its own in regulating inclusion and exclusion. Never before had religion in this sense been a network of decision premises. And never before did its own unity of reproduction become so dependent on interpretation, i.e., professional skill in handling distinctions.

This kind of self-regulation seems to require another semantic innovation. The old difference between sacred and profane, applied to places, occasions, persons, etc., had to be replaced with a difference that could be handled as a purely internal difference within the religious system itself, representing as it were the differences between those included in and those excluded from the religious system. This problem was solved by the distinctions between *salvation* and *damnation*, accessible to all kinds of clerical and private manipulation. This difference could be presented to the believer as the most important question of his life. It could, then, be conditioned by all kinds of secondary regulations. And even in the face of nonbelievers Pascal and others (Pascal and Jesuits!),[28] one could argue that it was not worthwhile to risk damnation even if one did not believe in it. The scheme could be handled as a totalizing device, including the whole world and even those excluded. At this level, the paradox became a suggested calculus of decision.[29] As a result of long debates the paradoxes surrounding salvation became more prominent in the late Middle Ages. And whereas tradition did maintain a simple inverse relation between certainty of salvation and fear in everyday life—the more certainty the less fear—the problem became exaggerated, and culminated in the issue of salvation itself: in its uncertainty.

Another area of problems relates to communication. A long process of doctrinal evolution has reduced the possibility of communication with the sacred to two forms: revelation and prayer.[30] The same process had intensified the communicative character of revelation and prayer and, thereby, gave rise to private concerns. When Japanese beat the gong, bow, and think of wishes in front of the temple, we don't know for sure whether this is intended as communication or not. The Christian prayer is intended as communication and therefore requires a sufficient distinctness of belief. Revelation, too, does not simply create states, consecrate places, destroy the evil, or interfere in some other way in worldly affairs. Again it is intentional communication, and this means freedom to accept or not to accept the message. Since God can and cannot reduce Himself to something visible (again a paradox!), He sent His Son to preach the gospel.

The result of this doctrinal evolution is differentiation. The specification of forms of communication between God and man leaves the relation of man and nature free for other concerns—be they economic, scientific, or aesthetic. All of these concerns retain a religious quality too, because God has created the world and given nature to man. But there is no communicative relation between man and nature.[31] This must have been a very difficult decision, possible only with religious support. Franciscus d'Assisi talks to animals. The way Petrarca sees nature almost becomes a new religion. Scientific experiments are styled as questioning nature. But actually nature remains silent as an object of pleasure and exploitation. It does not complain.

This stupendous and unique construction of theological doctrine was possible only on the basis of structural differentiation. Above all it presupposed a separation of political and religious roles and a certain "privatization" of religious concerns, already realized during the classical period of the Greek city-state.[32] This structural differentiation made it possible to think of membership in religious organizations as a matter of private choice and to begin to develop decision premises and rules of control that made it feasible to separate members and nonmembers without using other roles (e.g., citizenship) as a guideline. The decision to belong or not to belong to a certain religious collectivity became independent from other roles of the individual. The articulation of belief was necessary to orient this choice, and the paradoxical structure of belief (e.g., a man of lowest birth the Son of God) could symbolize the independence of this choice. It is one of the accidents of evolution that this condition lasted long enough for the consolidation of a belief system that could survive the abolition of its starting mechanism. The established church into which we are born did retain (with new meanings) ceremonies of enrollment and admittance (baptism) and, above all, the independence from other roles. Everybody can become a Christian: a son, a wife, a slave, a heathen of whatever complexion, and even a criminal.

There is a circular relation of reciprocal support between semantics and social structure which for a long time stabilized the result of an improbable evolution. However, we are recovering the improbability of the probable. The religious system evolved and it had to pay the penalty. The inherent improbability reappeared as a discrepancy between semantics and social structure and as a permanent incitation to reform. The Church did not live up to its own expectations. From the twelfth century, it became the object of more or less continuous claims for spiritual and organizational reforms, and it became hardened by accepting and rejecting reforms. This, too, contributed to differentiation. No other institution had a similar history. The differentiation of religion and politics became practically irreversible, and it became one of the main conditions for a new type of solution: for the differentiation of the mother church itself into several churches, sects, and denominations.[33]

At the same time, a new differentiation of religious and economic questions emerged. The religious system had to renounce any attempt to supervise and justify economic behavior—church policy in matters of usury and just prices having been its main foothold in divine economic consultancy[34]—and the economic system had to renounce any attempt to buy salvation. Both systems had to look for less immediate forms of mutual influence, respecting the autonomy of the other. Quite similar problems of structural differentiation came up in relation to areas of personal intimacy. The religious system had to withdraw from regulating the position of bodies engaged in sexual activities,[35] but it could stop all attempts to point to the woman as the way to salvation—from Schlegel's *Lucinde* to Claudel's *Soulier de satin*.[36]

Thus, evolution of religion is not simply a change of religious forms. The point is not simply the development of a clearer conceptualization of the paradox. It is differentiation in a more complicated sense. Evolution propels itself by changing *systems* which are at the same time the *environment for other systems*, forcing the latter to adapt or resist. This may amount to changing structures or retaining unchanged structures within a changed environment —but in both cases it amounts to a strengthening of differentiation. Under these pressures of social evolution structural differentiation seems to reinforce and extend functional specification, and the result is the functional differentiation of the whole society, the modern type of society we are all familiar with.[37]

Semantic and structural differentiation of religion leaves other areas of life without religious support. Their structures remain inherently paradoxical, if they cannot be reformulated in religious terms. Classical political economy, for example, defined its concept of labor, as the relation between man and nature. This contained obvious references to biblical tradition. But labor is no longer the consequence of original sin or an element within the religious dramaturgy of salvation. It is a natural necessity, even a "natural law." Thus, the paradox reenters the theoretical framework of political economy: the relation between man and nature is again a natural relation. Therefore endless controversies cropped up concerning the status of labor within the system of economic production and distribution, and any solution had to rely, if not on religious, on ideological deparadoxization.[38]

Today, religion survives as a functional subsystem of a functionally differentiated society. It has gained recognized autonomy at the cost of recognizing the autonomy of the other subsystems, i.e., secularization.[39] It represents the world within the world and society within society. Its paradox can be reformulated as the well-known paradox of set theory: it is a set that includes and excludes itself.[40]

The traditional way to deparadoxize this paradox has been "representation." The modern way seems to require a functional orientation. The "deparadoxization" (I am trying to find a linguistic correlate for the improbability of the probable) of the world becomes a job, and "calling God" becomes the solution of a problem. At the same time, we know how inadequate it is to treat religion in this way. We may ask whether a solution will be found at all for the problem of religion, or for that matter for the problem of meaning; and we may also ask whether our solutions, and particularly the solution we call modern society, can find their problem. We know of countermovements, the recent reactions of the Islamic religion against secularization being the most spectacular one.[41] But defining the modern way of life or Western style or capitalist society or secular rationality in negative terms and reacting to it by negating this negativity is in itself a very modern way of coping with problems and, as we well know, not a very successful one.

A less fundamental and more appropriate way would be to look for adequate theoretical descriptions of this very situation, not negating, but abstracting from the framework in which we experience modern life. We could, for example, start by revisiting the semantic and structural choices made by the system of religion as it was approaching modern times. We may ask:

1. Was it a good idea to strengthen, beginning with the Council of Trent and with Protestant "state churches," the *organizational infrastructure* of the religious system, to reduce its professionals to a status of functionaries, and to develop a hierarchical unification although this centralized power of programming and decision making proved not to be able to adapt religion to modern conditions.[42]

2. Was it a good idea to symbolize the paradox by a semantics of *invisibility*[43] which was, by the way, always known or felt to be unsatisfactory with respect to religious needs?[44]

3. Was it a good idea, this being perhaps the most important of all of these semantic changes, to drop the notion of *hell*,[45] to renounce terror and fear in religion, to present it as pure love and thereby lose the distinction between salvation and damnation, the only binary schematism specified for the religious system?[46]

It is easy to see that these and similar structural changes responded to the functional differentiation and to the increasing complexity of modern society. It is difficult to see any alternatives, and it would be presumptuous to say that this was all wrong. The point is that sociological theory and particularly systems theory offer a conceptual framework for describing such developments in more abstract terms, for example the distinction between society and organization as different types of social systems; the notion of semantic reformulations of paradoxes; or the notion of binary schematisms (good/bad, true/false, right/wrong, healthy/sick, salvation/damnation, to have or not to have property, etc.) as information-processing devices. Described in this way religion can be perceived as developing along lines that are partly typical and partly untypical of other functional subsystems of modern society. A reliance on organization is characteristic of the political system, focusing on the "state," but not of science. A reliance on formulas that bypass the original problem is typical of education,[47] but not of art, and perhaps of the economy, but not of the medical system. Trying to get along without any fundamental distinction, without any binary schematism, seems to be a unique experiment, characteristic of the religious system only. It looks as if the monotony of a loving God had to compensate for the diminishing importance of religion in everyday life,[48] and it seems that this reinforces the organizational differences between members and nonmembers of churches or denominations. And above all, abandoning the fundamental differences between salvation and damnation, materialized as heaven and hell, leads back to the fundamental paradox of self-referential unity.[49] Religion returns to its original problem.

In sociological terms this original problem is the problem of paradoxical self-reference. In religious terms it may be formulated —and formulation is already a kind of solution—as the problem of transcendence. In fact, the essence of the surviving religious traditions can be subsumed under this heading.[50] Seen from a metaperspective both formulations may have the same meaning.

Within the context of traditional religious formulation, transcendence is conceived as something *given*, an almighty power of creation and/or interference from the outside. In the eyes of an anthropologist or sociologist this is but the solution of a problem, transcendence being an imaginary *creation* of man to solve problems of meaning within the world. Each position can take account of the other. To the religious mind, sociologists, living without faith and in a state of sin and limited knowledge, have no chance to see the reality of transcendence. Maybe they took the wrong apple. As sociologists see it, religious people are faced with the problem of "latent functions." They cannot be aware of the functions of their belief because this would destroy the belief itself. They cannot believe in the function of their belief,[51] they cannot believe in "deparadoxization," and have to remain in the shadowy cave of everyday life. However, this may be but a battle of academic disciplines or intellectuals.[52] And this, again, may be but an exercise in self-reference, using contradictions to make one's own point. Why are we supposed to decide on this issue? To paraphrase Ranulph Glanville, the question of fundamentals is not necessarily a fundamental question.[53]

The main problem of contemporary religious practice might well be the problem of transcendental communication. For structural reasons our society discourages any attempt to communicate with partners in its environment. The universe has withdrawn into silence. But relations between God and man have to be communication—or what else?—yet cannot be communication.[54]

The Bible itself seems to react to its own increasing literacy. "Hearing" the voice of God had become a written text, a report about past events, and thus was no longer possible in the same sense. God had to send His Son to be audible. He did send Him *as his word.* Eo verbum quo filius. But this again became part of the same written book and will not be repeatable. Today this impossibility of communication is not only enforced by writing, it is reinforced by the structural development of the societal system. All communication reproduces society and remains a strictly internal operation. Moreover, only human beings can support the social network of communication. Communication with gods, like communication with pets, may be emotionally gratifying; but it operates, at least for observers, somewhat out of touch with reality— like "hearing voices." "Calling God" in public places amounts to strange behavior or to socially oriented communication, e.g., by bumper stickers. Our normal understanding of communication points to human receptors and all the refinements of awareness, and empathy makes this so much more unavoidable.

We can of course *say* that we *mean* something different by communication with God. But then, what do we mean? And can we, without stumbling over the paradox, *say* that we do not *mean* what we *say* knowing that others will not know what we mean when we say that we do not mean what we say?

We can renounce any attempt at active or passive transcendental communication. But then, we would admit that we have to rely on psychological and social resources or reinforcement of belief and would again be faced with the invisible God and the situation *etsi non daretur Deus*. We have churches. They are places where calling God, explaining His revelation (as if it were communication), and prayer is adequate and expected behavior. In sociological terms churches seem to cultivate countermores, depending for their success on being different. Religion may have become counteradaptive,[55] and this may be the very reason for its survival and for its recurrent revival as well. The Church itself, by now, may have become a carnival, i.e., the reversal of normal order.[56]

To propose this account may be sound sociological reasoning. And it would be good sociological theory, were it not for the fact that the function of religion refers to the constitutive paradox of the whole society as a self-referential system. On the one hand we can admit that enclosure of the paradox, counteradaptive behavior, preserving memory, and keeping a place where the unusual may become usual, the unbelievable believable, the improbable probable may be the solution; on the other hand, it is part of the functional perspective to look for functional equivalents and to keep asking the question whether and why we have to be satisfied with this sort of paradoxical solution.

Endnotes

1. By the way, to preserve "classical" authors and to need them is a relatively new phenomenon. Until the eighteenth century, the term "classical" simply meant: used in school classes.

2. See Clifford Geertz, "Religion as a Cultural System," in Michael Banton ed., *Anthropological Approaches to the Study of Religion* (London: Tavistock, 1966), pp. 1–46.

3. The research to which I refer has interdisciplinary relevance. It is therefore very heterogeneous. It includes: Robert P. Pula, "General Semantics as a General System Which Explicitly Includes the System Maker," in Donald E. Washburn and Dennis R. Smith, eds., *Coping with Increasing Complexity: Implications of General Semantics and General Systems Theory* (New York: Gordon and Breach, 1974), pp. 69–84 (based on Korzybski); Gotthard Günther, "Cybernetic Ontology and Transjunctional Operations," in Gotthard Günther, *Beiträge zur Grundlegung einer operationsfähigen Dialektik* (Hamburg: Meiner, 1976), 1:249–328; Edgar Morin, *La Méthode* (Paris: Seuil, vol. 1, 1977; vol. 2, 1980); Yves Barel, *Le paradoxe et le système: Essai sur le fantastique social* (Grenoble: Presses Universitaires, 1979); Francisco Varela, *Principles of Biological Autonomy* (New York: North-Holland, 1979); Humberto R. Maturana and Francisco J. Varela, *Autopoiesis and Cognition: The Realization of the Living* (Dordrecht: Reidel, 1980); Erich Jantsch, *The Self-Organizing Universe: Scientific and Human Implications of the Emerging Paradigm of Evolution* (Oxford: Pergamon, 1980); Milan Zeleny, ed., *Autopoiesis: A Theory of Living Organization* (New York: North-Holland, 1981); Heinz von Foerster, *Observing Systems* (Seaside, Calif.: 1981).

4. For an extensive discussion see Niklas Luhmann, *Soziale Systeme: Grundriß einer allgemeinen Theorie*, (Frankfurt: Suhrkamp, 1984).

5. Since this innovation is largely due to Humberto Maturana, the best formulations might well be his *ipsissima verba:* "We maintain that there are systems that are defined as unities as networks of production of components

that (1) recursively, through their interactions, generate and realize the network that produces them; and (2) constitute, in the space in which they exist, the boundaries of these networks as components that participate in the realization of the network." And for a further elucidation: "An autopoietic system is defined as a unity through relations of production of components, not through the components that compose it whichever they may be. An autopoietic system is defined as a unity through relations of form (relations of relations), not through relations of energy transformation. An autopoietic system is defined as a unity through the specification of a medium in its realization as an autonomous entity, not through relations with a medium that determines its extension of boundaries." Maturana, "Autopoiesis," in Zeleny, *Autopoiesis*, pp. 21–33, 21 and 29f.

6. I avoid the terms "communicative acts" or "speech acts" and I avoid any reference to "action theory" as the basic framework. Units of communication are not simply acts or actions. They are unified (within the social system) by a synthesis of information, utterance, and understanding. (The German language would allow me to distinguish between *Kommunikation* in this sense and *Mitteilung* as a communicative act.)

7. See Niklas Luhmann, "Sinn als Grundbegriff der Soziologie," in Jürgen Habermas and Niklas Luhmann, *Theorie der Gesellschaft oder Sozialtechnologie: Was leistet die Systemforschung?* (Frankfurt: Suhrkamp, 1971), pp. 25–100 (chapter 2 in this volume).

8. Edgar Morin, *La Méthode* (Paris: Seuil, 1977), 1:201.

9. See Gregory Bateson, *Steps to an Ecology of Mind* (New York: Ballantine, 1971), pp. 405ff., 417ff.

10. I use this term in its logical and theological sense, defined by the negation of impossibility and the negation of necessity. The understanding of "contingent" as "dependent on" is only a restricted version, depending itself on the theological interpretation of contingency as created by and depending on the will of God.

11. This is, of course, a sociological reconstruction of a famous theological problem. See, for some aspects of a copious literature, Thomas B. Wright, "Necessary and Contingent Being in St. Thomas," *The New Scholasticism* (1951), 25:439–466; Guy Jalbert, *Nécessité et contingence chez saint Thomas d'Aquin et chez ses prédécesseurs* (Ottowa: 1961); Innocenzo d'Arenzano, "Necessità e contingenza nell'agire della natura secondo San Tommaso," *Divus Thomas* (1961), vol. 38, 3d series, pp. 27–69; and Celestino Solagurén, "Contingencia y creación en la filosofía de Duns Escoto," *Verdad y Vida* (1966), 24:55–100.

12. See Georg Simmel, *Die Religion* (Frankfurt: Rütten and Loening, 1906). See also his "Zur Soziologie der Religion," *Neue Deutsche Rundschau* (1898), 9:111–123.

13. See Roy A. Rappaport, "The Sacred in Human Evolution," *Annual Review of Ecology and Systematics* (1971), 2:33–44; and Roy A. Rappaport, "Ritual, Sanctity and Cybernetics," *American Anthropologist* (1971), 73:59–76.

14. See Roy A. Rappaport, "Maladaptation in Social Systems," in J. Friedman and M. J. Rowlands, eds., *The Evolution of Social Systems* (Pittsburgh, Pa.: University of Pennsylvania Press, 1978), pp. 49–71.

15. See Paul Valéry, "Eupalinos ou l'architecte," in Paul Valéry, *Œuvres* (Paris: éd. de la Pléiade, 1960), 2:79–147.

16. See Robert N. Bellah, "Religious Evolution," *American Sociological Review* (1964), 29:358–374.

17. See Marion Blute, "Sociocultural Evolutionism: An Untried Theory," *Behavioral Science* (1979), 24:46–59.

18. See Varela, *Principles of Biological Autonomy*, pp. 37ff.; Gerhard Roth, "Conditions of Evolution and Adaptation in Organisms as Autopoietic Systems," in D. Mossakowski and G. Roth, eds., *Environmental Adaptation and Evolution* (Stuttgart and New York: Fischer, 1982), pp. 37–48. I know of no discussion of this point in sociology.

19. See Eric A. Havelock, *The Literate Revolution in Greece and Its Cultural Consequences* (Princeton, N.J.: Princeton University Press, 1982).

20. See Walter J. Ong, *The Presence of the Word: Some Prolegomena for Cultural and Religious History* (New Haven, Conn.: Yale University Press, 1967); Walter J. Ong, "Communications Media and the State of Theology," *Cross Currents* (1969), 19:462–480.

21. See Walter J. Ong, *Interfaces of the Word: Studies in the Evolution of Consciousness and Culture* (Ithaca, N.Y.: Cornell University Press, 1977), pp. 20f.

22. Ong, in *The Presence of the Word*, p. 190f., even suggests that God Himself chose this historical moment where orality was still important but the alphabet already available—"the precise time when psychological structures assured that this entrance would have greatest opportunity to endure and flower."

23. See Samuel H. Monk, *The Sublime: A Study of Critical Theories in Eighteenth-Century England*, 2d ed. (Ann Arbor, Mich.: University of Michigan Press, 1960).

24. See Eric A. Havelock, *Preface to Plato* (Cambridge, Mass.: Belknap, 1963).

25. See Jack Goody and Ian Watt, "The Consequences of Literacy," *Comparative Studies in Society and History* (1963), 5:304–345.

26. See Anthony, Earl of Shaftesbury, "Miscellaneous Reflections," in his *Characteristicks of Men, Manners, Opinions, Times* (1714; reprint Farnborough, England: Gregg, 1968), 3:81.

27. See Talcott Parsons, "Evolutionary Universals in Society," *American Sociological Review* (1964), 29:339–357.

28. For a Jesuit example of the famous wager of Pascal see Pierre de Villiers, *Pensées et reflexions sur les égaremens des hommes dans la voye du salut*, 3d ed. (Paris: Collombat, 1700), 1:204f.

29. There are many secondary paradoxes associated with salvation, for example the idea that the most external sign is given by God as the most reliable warrant of internal certainty: *verbum solum habemus*; or later, as Max Weber would have it: business only!

30. This statement relates to preaching and to popular religion. Theology, in a technical sense, can of course avoid or circumvent the concept of communication with God, using concepts like *invocatio* or *evocatio* or reducing this difficult operation to: calling God by His name.

31. The substitute is, of course, labor.

32. For the very advanced state of structural differentiation and religious privatization in Athens, see S. C. Humphreys, "Evolution and History: Approaches to the Study of Structural Differentiation," in J. Friedman and M. J. Rowlands, eds., *The Evolution of Social Systems*, (Pittsburgh: University of Pennsylvania Press, 1978), pp. 341–371.

33. One has to look at the very complicated relations between religious reform movements and the emerging territorial states in the fifteenth century to see this point. The new forms of symbiosis of church and state that emerged on the basis of differentiation were the precondition, not the result, of the Protestant Reformation.

34. See Raymond de Roover, "The Concept of Just Price: Theory and Economic Policy," *Journal of Economic History* (1958), 18:418–434; and Marjorie Grice-Hutchinson, *Early Economic Thought in Spain 1170–1740* (London: Allen and Unwin, 1978). See also Benjamin Nelson, *The Idea of Usury: From Tribal Brotherhood to Universal Otherhood*, 2d ed. (Chicago: 1969); Robert P. Malorey, "Usury in Greek, Roman and Rabbinic Thought," *Tradition* (1971), 27:79–109.

35. See Jean-Louis Flandrin, *Familles: Parenté, maison, sexualité dans l'ancienne société* (Paris: Hachette, 1976); and "Uomo e donna nel letto coniugale—norme e comportamenti: La teologia morale di fronte alla sessualità," *Prometeo* (1983), 1(4):44–49.

36. See Jochen Hörisch, *Gott, Geld und Glück: Zur Logik der Liebe in den Bildungsromanen Goethes, Kellers and Thomas Manns* (Frankfurt: Suhrkamp, 1983).

37. See Niklas Luhmann, *The Differentiation of Society* (New York: Columbia University Press, 1982), esp. pp. 229ff.

38. We can see this consulting a text from Thomas Hodgskin, *Popular Political Economy* (London: Tait, 1827; reprint New York: Kelley, 1966), pp. 28ff.: "It is a law of our being, that we must eat bread by the sweat of our brow; but it is reciprocally a law of the external world, that it shall give bread for our labour, and give it only for labour. Thus we see that the world, every part of which is regulated by unalterable laws, is adapted to man, and man to the world." (p. 28). Hodgskin admits "that men have, to a certain degree, the power of throwing the necessity to labour off their own shoulders; as they may alter the direction of the influence of gravity" (p. 30). But "every long-continued attempt in one class of men to escape from the necessity of labour imposed on our race. . . . is a violation of a natural law which never passes unpublished" (p. 30). This is in part biblical political economy, but it leads, by dropping the religious reference, to a natural law of class politics.

39. See Niklas Luhmann, *Die Funktion der Religion* (Frankfurt: Suhrkamp, 1977), pp. 225ff.

40. See, in relation to problems of "enlightenment," Marco M. Olivetti, "Sich in seinem Namen versammeln: Kirche als Gottesnennung," in Bernhard Casper, ed., *Gott nennen: Phänomenologische Zugänge* (Freiburg and Munich: Alber, 1981), pp. 189–217.

41. With respect to "functional differentiation" as societal background of defensive aggressiveness see Bassam Tibi, "Islam and Secularization: Religion and the Functional Differentiation of the Social System," *Archiv für Rechts- und Sozialphilosophie* (1981), 66:207–221; and Bassam Tibi, *Die Krise des modernen Islam: Eine vorindustrielle Kultur im wissenschaftlich-technischen Zeitalter* (Munich: Beck, 1981).

42. For critical views, see Michel de Certeau, "Du système religieux à l'éthique des Lumières (17e–18e s.): La Formalité des Pratiques," *Ricerche di Storia Sociale e Religiosa* (1972), 1(2):31–94; and Franz-Xaver Kaufmann, *Kirche begreifen: Analysen und Thesen zur gesellschaftlichen Verfassung des Christentums* (Freiburg/Brsg.: Herder, 1979). See also Niklas Luhmann, "Die Organisierbarkeit

von Religionen und Kirchen," in Jakobus Wössner, ed., *Religion im Umbruch* (Stuttgart: Enke, 1972), pp. 245–285.

43. In the eighteenth century the metaphor of *invisibility* or *hiddenness* was used in many different contexts to solve the paradoxes of order (e.g., the hidden order of order and disorder, the unity of multiplicity, etc.) The well-known "invisible hand" referred to by Joseph Glanvill, Adam Smith, and others is only one case in point, formulating the expectation of progress in functionally differentiated subsystems (science, economy). For a cosmological argument see Johann Heinrich Lambert, *Cosmologische Briefe über die Einrichtung des Weltbaues* (Augsburg: Klett, 1761), p. 116: "Die Unordnung in der Welt ist nur scheinbar, und wo sie am größten zu seyn scheint, da ist die wahre Ordnung noch weit herrlicher, uns aber nur mehr verborgen."

44. See a scene in Louis-Sébastien Mercier, *L'homme sauvage, histoire traduite de . . .* (Paris: Duchesne, 1767), with the feeble comfort: "un jour nous le connaîtrons" (p. 119). See also, for the seventeenth-century, Lucien Goldmann, *Le dieu caché* (Paris: 1956). The point is no longer *hearing* the voice of God (Gen. 3:8, 7:1, 8:15, 22:1, 31:11, etc.) but *seing* something; and only the surface of things is visible behind which something else may be hidden.

45. Hell being but an intellectual instrument of priests to terrify and dominate the people, runs the argument. See Pierre Cuppé, *Le ciel ouvert à tous les hommes, ou traité théologique* (Paris: 1768); anon., (=Dom Nicolas Louis), *Le ciel ouvert à tout l'univers* (1782). See also William Blake, "The Marriage of Heaven and Hell" (1790–1793), in William Blake, *Complete Writings*, ed. Geoffrey Keynes (London: Oxford University Press, 1969), pp. 148–158, with a different view concerning the *function* of the difference: "Without Contraries is no Progression. Attraction and Repulsion, Reason and Energy, Love and Hate, are necessary to Human existence. From these contraries spring what the religious call Good & Evil. Good is the passive that obeys Reason. Evil is the active springing from Energy. Good is Heaven. Evil is Hell."

46. One of many critical observations concerned questions of space in hell. After some time it must have become a terribly crowded place without enough fire for all because it had to preserve bodies; whereas in heaven the soul could be thought of as an infinitely compressible substance.

47. Particularly for Germany, but not for France. See Niklas Luhmann and Karl Eberhard Schorr, *Reflexionsprobleme im Erziehungssystem* (Stuttgart: Klett-Cotta, 1979), and Jürgen Schriewer, "Pädagogik—ein deutsches Syndrom? Universitäre Erziehungswissenschaft im deutsch-französischen Vergleich," *Zeitschrift für Pädagogik* (1983), 29:359–389.

48. Cuppé and Louis (see note 45) could at least imagine a system of gradation in heaven to replace the crude difference between heaven and hell with a careerlike distinction between better and worse positions.

49. See Barel, *Le paradoxe et le système*, pp. 89f.

50. See Edwin Dowdy, ed., *Ways of Transcendence: Insights from Major Religions and Modern Thought* (Bedford Park, South Australia: The Australian Association for the Study of Religion, 1982).

51. An argument, by the way, that is much older than sociology as an academic discipline. See Peter Villaume, *Über das Verhältnis der Religion zur Moral und zum Staate* (Libau: Friedrich, 1791). The argument runs: functional explanations have no access to the truth of what they explain. They have to take everything as "opinion."

52. Including me. See Niklas Luhmann and Wolfhart Pannenberg, "Die Allgemeingültigkeit der Religion," *Evangelische Kommentare* (1978), 11:350–357. See also Frithard Scholz, *Freiheit als Indifferenz: Alteuropäische Probleme mit der Systemtheorie Niklas Luhmanns* (Frankfurt: Suhrkamp, 1982).

53. See Ranulph Glanville, "The Nature of Fundamentals, Applied to the Fundamentals of Nature," in George J. Klir, ed., *Applied General Systems Research: Recent Developments and Trends* (New York: Plenum Press, 1978), pp. 401–409. See also another piece of British wisdom: "in reality, *profound Thinking* is many times the Cause of *shallow Thought.*" Earl of Shaftesbury, "Miscellaneous Reflections," p. 226.

54. See again Blake, "The Marriage of Heave and Hell," p. 153: "The Prophets Isaiah and Ezekiel dined with me, and I asked them how they dared so roundly to assert that God spoke to them; and whether they did not think at the time that they would be misunderstood; and so be the cause of imposition. Isaiah answer'd: 'I saw no God, nor heard any, in a finite organical perception; but my senses discover'd the infinite in every thing, and as I was then perswaded, & remain confirm'd, that the voice of honest indignation is the voice of God, I cared not for consequences but wrote.' "

55. "Counteradaptive" as the evolutionary result of adaptive advances in earlier times. See Anthony Wilden, *System and Structure: Essays in Communication and Exchange*, 2d ed. (London: Tavistock, 1980), pp. 205ff. See also the reference to "maladaptive" in note 14.

56. See Mikhail Bakhtin, *Rabelais and His World* (Cambridge, Mass.: MIT Press, 1968). See also David Gross, "Culture and Negativity: Notes Toward a Theory of the Carnival," *Telos* (1978), 36:127–132.

9.
The "State" of the Political System

Today the theory of the state and the theory of political systems belong to different realms of scientific discourse. Political scientists and sociologists are used to speaking of political systems. Within the legal discourse, at least in Europe, the notion of the state is preferred—partly because it is used by the law itself and partly because it retains a tradition of speaking about the central focus of all political activities. Moreover, if we speak of the political system, our subject is a subsystem of the society on the same level as the economic system, the system of science, the educational system, etc. The semantics of the state, on the other hand, is based upon a distinction between state and society that suggests that the state exists outside of the society. The state, then, is seen as a legal person or as a collective actor being different from the network of private needs and private interests which is the stuff out of which the society is built.[1] To lawyers this distinction seems to be a condition of legal imputation; for how can you attribute an action to the state if the state does not exist as an entity per se with freedom of will outside of the casual interconnections of the society?

From the point of view of the social sciences this is at best a legal fiction. In social reality, the state seems to be nothing more and nothing less than the public bureaucracy, including parliaments and eventually courts, schools, and public services, but excluding political parties, interest groups, social movements pursuing political goals, and even the electorate. These other social collectivities and groups give "inputs" to the state. They belong to the political system if, and only if, they organize demands and pressures addressed at the state as the center of political power.

Given these distinct approaches to political and legal affairs, the formulation of my title, "The 'State' of the Political System," becomes ambiguous and difficult to understand it. It revives the old way of using the term *state*, customary in Europe from the fifteenth to the eighteenth centuries.[2] State, like Latin status, simply meant the actual conditions and present situation of political power, including financial and military means, external and internal relations, good counsel and good luck.[3] However, my intention is not to sound old-fashioned or to use the term *state* in its colloquial, nonpolitical sense. I want to crosscut the two ways of speaking about politics, the two languages, the two discourses, because I see no reason to keep them separate.

The term *political system* has the advantage of being connected with the very general and powerful framework of systems theory.[4] It draws on analytic resources that are developing rapidly without particular regard to the special field of politics. Its most recent paradigmatic shift points in the direction of a general theory of self-referential systems, including social systems and, again as a special case, societies.[5] Within this general framework the concept of the political system can be used to describe the (degree of) differentiation of politics from other social concerns. Self-reference is a very general principle of system building. Differentiation can be observed as a concrete historical process. Both, however, go together. Differentiation is differentiation of self-referential systems within the broad context of social evolution.

Starting with the concept of the political system we can see the formula *state* as a self-description of the political system. Complex systems are a result of evolution. They are not able to use their own complexity as means for the goals of the system because they cannot introduce their own complexity into the system. This would only mirror and multiply their original complexity and would make them hypercomplex. All self-awareness and all communication about the system within the system needs self-simplifying devices, i.e., identities. For the political system this function is fulfilled by the state.

The state, then, is not a subsystem of the political system. It is not the public bureaucracy. It is not only the legal fiction of a collective person to which decisions are attributed.[6] It is the political system reintroduced into the political system as a point of reference for political action.

This theoretical position has important advantages. It provides a new access to the history of the formula *state* and to its semantic career. It makes clear the sense in which the constitutional state is a special type of political system. It opens questions with respect to the welfare state and it makes it possible to reanalyze the paradoxes of "collectively binding decision making," the core function of politics.

The theoretical framework of social evolution as increasing differentiation of self-referential systems generates the hypothesis that the subsystems of society, by differentiation, become increasingly complex. They will lose their capacity to use their own complexity and will break down if no strategic adaptations are developed. They then need adequate self-descriptions. We therefore have to suppose a relation between (1) increasing differentiation, (2) increasing complexity of differentiated systems, and (3) the development of self-simplifying devices that make it possible to use the system as a premise of its own operations. In fact, the history of the state formula seems to verify this hypothesis.

Within the European society of the late Middle Ages political units became more independent of economic households. They developed from personal collectivities into territorial units. They became, already in the century of the conciliar movement before the Reformation, adversaries and partners of church policies and thereby differentiated from the religious system. As a result, the political apparatus, which dominated countries within fixed boundaries and tried to maintain absolute (i.e., independent) and sovereign power, became increasingly complex.[7] It needed all the more a representation of its own unity.[8] The monarch was considered as an indispensable symbol—even by Hegel. He was not necessarily regarded as a powerful person, but as the representation of the unity of a complexity by the individuality of a person.

There are a lot of controversies about the beginnings of the modern state: late Middle Ages or only sixteenth century? In any case, the breakthrough at the semantic level, the conceptual reformulation, comes relatively late. The discussion of a special *ragion di stato* did produce more smoke than fire. Machiavelli himself did not use the term. Essentially, the question was whether or not special circumstances and special responsibilities or common utility give a special right to derogate the law and to deviate from accepted standards of morality, and in addition: how to construct derogability,[9] within or outside of natural law. This problem of derogation had medieval roots.[10] The difficulty was to formulate limiting conditions for what later came to be called *ius eminens*. This made it compelling to separate the private interests of the prince and care for public welfare. It did not, at least not immediately, suggest a new concept for the identity of the political system.

Not these legal moralistic questions but rather new problems of complexity did require the formulation. As early as the eighteenth century complexity became the latent problem of political theories,[11] and at the same time the concept of the state changed its meaning from colloquial to conceptual significance. It had been an attribute of something else, describing the changing conditions of its maintenance. It became a term for an object of its own, a term for the unity of the multiplicity of political goals and activities. After this semantic change it was possible to say that the monarch is the first servant of the state. The eighteenth century did still retain the old notion of the "civil (i.e., political) society" and therefore had to speak of "political" issues in a very broad sense, in spite of, and together with, the new semantics of the state.[12] The nineteenth century experienced the consequences, and politics became defined as everything that relates to the state.

Thus, the notion of the state became an historical concept. By this term I mean not only a concept that has been used in history. Historical concepts are concepts that make a difference in history. They thereby move history. This historical difference, then, becomes part of their meaning. The state is, in fact, the modern state.[13] Any definition that delineates the meaning of the state—be it in distinction to the ruling dynasty or in distinction to the society, be it as legal person or as unified power of decision about the use of force in a given territory—gives only part of its meaning. To completely understand the meaning of the state requires that one understand the historical situation in which something like "the state" became necessary, that is, the situation in which a formula for the self-description of the system had to be invented, given the only alternative: that the political system will not operate at the required level.[14]

A self-description formulates the identity of the system. However, identity can be identified only by establishing a difference. The ancient formula of the civil society (with *politeia*/Polizey as its defining focus) presupposed the distinction from uncivilized, barbarian conditions. The eighteenth century renewed this difference by distinguishing between natural states and civilization (and this difference was being made by the introduction of division of labor).

"STATE" OF THE POLITICAL SYSTEM   169

Thus, European society could describe itself as progressing from natural to civilized states—whatever the costs of losing natural freedom, natural equality, natural morals, and natural unrepressed sexuality.

These descriptions had always been self-descriptions of the total society. The idea of the state for the first time introduced a quite new semantic constellation. The state was no longer thought of as the political society itself, but was defined by a distinction between state and society, roughly equal to the distinction between force and property.[15] Compared with the traditional semantics of civil society it was no longer an external difference but an internal difference, a social one that was used to describe the political system.

These semantic shifts prepared the ground for what came to be called the constitutional state (*Verfassungsstaat*). The constitution was supposed to perform the miracle of self-limitation of sovereign power. In this sense it was a paradoxical institution, combining within one legal instrument the unlimited and the limited. Since the monarch had to "give" the constitution, it was perceived not as a logical impossibility but as the outcome of political battle. The constitution itself became a political act, defining the possibilities and limitations of politics. The constitution was perceived as an attribute of the state, the state being an entity distinguishable by its constitution. Both ideas, state and constitution, reinforced each other. Apparently logical obstacles were not necessarily practical obstacles as the political resistance was broken. The result was the so-called "democratic" organization of political power.

This is well-known history. However, at the level of our analysis, we still have questions. To put it more simply: how was it possible to use the state to reorganize the political system? What kind of internal structure of this self-description generated the complex framework of constitutional provisions, including basic human rights, the division of powers, legitimate opposition, and public elections?

The answer to this question can be given by using the logical concept of reentry.[16] A distinction made to indicate a form, different from something else, reenters the form. The outside, as the side from which the distinction is supposed to be seen, becomes a premise of internal operations. A system, using reentry, can observe and describe itself. It can process information by taking the distinction between itself and the environment as its guideline. It not only knows the difference between itself and the environment (this every living system is supposed to do), but can control this difference as well by using the identity of the difference as a distinction within the system. In this sense the difference between state and society became part of the constitutional law and a premise for its interpretation. The political system, therefore, operates simultaneously at two logical levels; it operates as a paradoxical system, and the constitutional state is the formula for this operational paradox.[17] The paradox of self-limitation is replaced by the paradox of the reentry of the difference into the differentiated.

The Continental, and particularly the German, doctrine of the state has prevented this kind of analysis by mystifying the paradox, imitating, of course, theology.[18] The state was described as a real actor, as a collective individual, as a spiritual unity demanding moral participation and obedience. And in fact: how could one admit the paradox without granting the freedom to act as one pleases? A paradox has to be deparadoxized, if I may say so, to become a guideline for the operation of the system, and the constitution of the state is the means by which this requirement can be fulfilled.

As political evolution moves on, the state formula moves from *constitutional state* to *welfare state*.

This does not mean that the state loses its constitution. It certainly does not mean that it could do without it. But it means that new problems arise which cannot be solved by the legal norms of the constitution.

These problems too could be conceived of as offsprings of the paradoxical identity of the system and its self-description, and in this case as well we need a distinction between superficial and deep-structure descriptions. The constitutional state has its legal facade and its problem of reentry, which maintains that facade. The welfare state transforms this relation and gives it a new form.

The usual description of the welfare state refers to an historical process of increasing social engagements and activities. The state increasingly accepts responsibility for the solution of social problems. This leads to increasing financial burdens, to bureaucratization and legalization, and to an increasing dependency of everyday life upon state-controlled decisions. Thus, describing the state as a welfare state focuses upon a phenomenon of positive and negative growth and, since growth cannot continue infinitely, to a built-in crisis. The welfare state is a state in crisis, we would almost say: a state that wills its own crisis and may then use it as an occasion for continuous change of governments.

This, however, is only the first-level description. It may guide the normal self-observation of the political system (corresponding to the application of constitutional law of the constitutional state), but it gives no access to the hidden paradox of the system. Continuing growth is not paradox, it is only—impossible. The latent problem seems rather to be that together with new activities and new responsibilities the impact of the political system upon its social, human, and even physical environment increases. This has the effect that again and again new problems spring up from old solutions. As the state's share in problems and solutions grows, the environment overwhelms the system with new problems, which are in fact consequences of previous policies. Programs are put into operation with the best intentions but with unforeseeable, "counterintuitive" consequences. These side effects are fed back into the system as new issues for which responsibility cannot be denied. In this way things usually become more difficult, because the state tends to retain its previous goals, but has to cope with additional problems. Whereas the constitutional state could rely to large extent on the mechanisms of negative feedback, eliminating deviations from the law or eventually adapting the law by a slow process of juridical change, the welfare state has to cope with positive feedback, with increasing deviation as the very structure of its own policies. There is a close similarity to the flight of locusts: it can be stopped only by exhaustion.[19]

If the new paradox is that solutions create problems because problems create solutions; time becomes the critical variable, which can also mean that "saving" (i.e., gaining) time and avoiding decisions for the time being becomes the core virtue of politics. And consequently, it is not he who has the competence and power of final decisions who is sovereign, but he who has the possibility of avoiding situations in which he has no further alternative than to make a certain decision and to use his power.

Since a paradox is the logical equivalent of self-reference, we have to expect that social evolution can never avoid the paradoxical constitution of systems, but can only transform and update the ways in which a system tackles its own paradoxical identity. The same holds true if we look at the function of politics. It can be defined as providing for the continuing possibility of collectively binding decision making.[20] This formula refers not only to the possibility of enforcing the sovereign will upon others. Since this will cannot be thought of as existing outside of society (as the will of something like God), collectively binding decision making implies binding the decision maker himself. He then has to lose, to spend and to preserve his capacity to decide upon issues and to change, if convenient, his opinion.

This problem can be "solved" by creating artificial complexities, by introducing further distinctions, and by stating conditions under which decisions may be changed or respectively not changed. Since these conditions have to be introduced by decision—albeit by deciding about a constitution—the problem repeats itself on a higher level. No fixed point, no *Sittengesetz*, no principle of justice can be assumed as given and as framing the arbitrary will of the sovereign power. And even if this were possible it would, as thinkers of the eighteenth century knew well,[21] impose not enough constraint on the sovereign. Only complexity as such can help.

Once differentiated from its social environment with respect to its function the political system builds up complexity. The fundamental paradox seems to operate like an autocatalytical device— the paradox at least stays during all changes—and the evolution tends to eliminate disturbing complexity. Seen from this viewpoint, the state is important and, until today, is an irreplaceable evolutionary universal, which makes it possible to control relatively high complexity and to articulate the conditions that restrict the process of defeating and reestablishing, dissolving and reintroducing bindingness.

The managing of a system by a part of the same system is, of course, a very general problem.[22] It requires recursive solutions and it implies, above all, the capacity of self-observation on at least two levels: on a level of the total (managed) system, and on the level of the managing part-system. Formulating these problems— and they are classical problems of political theory—in terms of a theory of self-referential systems adds the insight that there are no other solutions compatible with system differentiation and system autonomy. On the whole, it seems as if evolution didn't take the time to produce a world according to logical and mathematical order. At least, it did use the stimulating power of paradoxes. The political system shares this way of producing order via tackling its paradox, and there seems no way to correct this kind of destiny after the fact. The loss of paradise was no accident.

Endnotes

1. See Ernst-Wolfgang Böckenförde, ed. *Staat und Gesellschaft* (Darmstadt; Wissenschaftliche Buchgesellschaft, 1976).
2. See Paul-Ludwig Weihnacht, *Staat: Studien zur Bedeutungsgeschichte* (Berlin: Duncker and Humboldt, 1968); and Wolfgang Mager, *Zur Entstehung des modernen Staatenbegriffs* (Wiesbaden; Steiner, 1970).
3. A reversed order, beginning with good luck, might be more appropriate, suggests Justus Lipsius in *Politicorum sive civilis doctrinae libri sex* (Leiden: 1589). See Karl-Georg Faber, "Macht, Gewalt," in Otto Brunner et al. eds., *Geschichtliche Grundbegriffe: Historisches Lexikon zur politisch-sozialen Sprache in Deutschland* (Stuttgart; Klett-Cotta, 1982), 3:875f.
4. See Niklas Luhmann, *The Differentiation of Society* (New York: Columbia University Press, 1982), pp. 138ff.; and Niklas Luhmann, *Politische Theorie in Wohlfahrtsstaat* (Munich: Olzog, 1981).
5. See Niklas Luhmann, *Soziale Systeme: Grundriss einer allgemeinen Theorie* (Frankfurt; Suhrkamp, 1984).
6. See Hans Kelsen, *General Theory of Law and State* (New York; 1961).
7. For a short outline see Gianfranco Poggi, *The Development of the Modern State: A Sociological Introduction* (Stanford, Calif.: Stanford University Press, 1978).
8. *Repraesentatio identitatis*, a term being invented by the conciliar movement of the fifteenth century to indicate a function that does not necessarily have to be fulfilled by the pope.
9. See Rodolfo de Mattei, "Il problema della deroga e la 'Ragion di Stato,'" in Enrico Castelli, ed., *Christianesimo e Ragion di Stato* (Rome and Milan: Bocca, 1953).
10. See A. Bonucci, *La derogibilità del diritto naturale nella scolastica* (Perugia: Bartelli, 1906).
11. See Harlan Wilson, "Complexity as a Theoretical Problem: Wider Perspectives in Political Theory," in R. La Porte Todd, ed., *Organized Social Complexity: Challenge to Politics and Policy* (Princeton, N.J.; Princeton University Press, 1975), pp. 281–331.
12. See Manfred Riedel, "Gesellschaft, bürgerliche," in: *Geschichtliche Grundbegriffe* (Stuttgart: Klett, 1975), 2:719–800 (738ff.).
13. Or "the modern states," as it has been referred to around 1800. The collective singular was introduced later. See Stephan Skalweit, *Der "moderne Staat": Ein historischer Begriff und seine Problematik* (Opladen: 1975).
14. This, of course, is a real possibility and even the probable development. Evolution proceeds by exception. Thus, when I said that concepts move history, I didn't fall back on a Hegelian type of *Geistesgeschichte*. The point is that semantic developments are necessary correlates of developments of social structures, and that in both cases evolution is a kind of counterintuitive movement.
15. In fact, the distinction between property and force was introduced earlier, giving substance and preparing the ground for the more abstract distinction between society and state. See, for one of many examples, François Véron

de Forbonnais, *Principes et observations oeconomiques* (Amsterdam; Revy, 1767), pp. 1ff. Besides, the differentiation of property and force itself was seen as a result of evolution: "Dans les temps anciens la richesse territoriale et la puissance étoient sinonimes et réunies dans le fait." (Forbonnais, *Principes*, p. 11, note a).

16. See George Spencer Brown, *Laws of Form*, 2d ed. (London: Allen and Unwin, 1971), pp. 69ff.

17. For a general introduction to related problems see Yves Barel, *Le paradoxe et le système: Essai sur le fantastique social* (Grenoble: Presses Universitaires, 1979).

18. See Carl Schmitt, *Politische Theologie: Vier Kapitel zu der Lehre von der Souveränität* (Munich: Duncker and Humblodt, 1922), and his less-known *Römischer Katholizismus und politische Form* (Munich: Theatiner Verlag, 1925), which looks back at this tradition from the "end of the state."

19. See T. Weis-Fogh, "An Aerodynamic Sense Organ Stimulating and Regulating the Flight in Locusts," *Nature* (1949), 164:873–874.

20. I see no serious alternative to this definition with the exception of formulations that have already deparadoxized the issue—say: reference to common interest or public goods or to the distinction (without reentry?) between friends and enemies.

21. So, for example, the physiocrats putting their stake on the constraints of an order of property. See Paul-Pierre Le Mercier de La Rivière, *De l'ordre naturel et essentiel des sociétés politiques* (London and Paris: 1767). But also, from a different point of view, Simon-Nicolas-Henri Linguet, *Théorie des loix civiles, ou Principes fondamentaux de la société*, 2 vols. (London: 1767).

22. For a sufficiently general statement see Heinz von Foerster, "The Curious Behavior of Complex Systems: Lessons from Biology," in Harold A. Linstone and W. H. Clive Simmonds, eds., *Futures Research: New Directions* (Reading, Mass.: Addison-Wesley, 1977), pp.104–113 (110ff.).

10.
The World Society as a Social System

The Concept of Society

Within the European tradition, a very general notion of society survived from the time of Aristotle until about 1800. The concept of society (*koinonoía, societas*) was almost identical with what we would call social system. The *encompassing* system was seen as a *special* case, namely as the political society (*koinonoía politiké, societas civilis*).[1] This conceptualization lost its significance with the emerging development of the modern state and of an industrialized economy. The old tradition cannot be revived.[2] It has, however, never been replaced with an adequate theoretical framework. There are attempts to change the dominant position of politics and to put economy or culture in its place. Such theories use a part of the reality of social life to represent the whole. Without giving sufficient reasons, economic or cultural or again political processes are postulated as the basic phenomenon. But the theory of these basic processes can claim only an historical and relative validity, since these processes are themselves part of sociocultural evolution.

General systems theory offers a new approach. At first sight, it looks like Aristotelian theory. A general notion of the social system is used to define the *encompassing* system as a *special* case of social systems. The content, however, has changed. Systems theory does not refer to the city or the state in order to characterize the special features of the encompassing system. Our society is too highly

This article was originally published in the *International Journal of General Systems* (April 1982).

differentiated for this kind of design. Instead, systems theory uses systems analysis to disclose the structures and processes that characterize the societal system—"the most important of all social systems which includes all others."[3]

Moreover, to conceive of societies as social systems excludes the traditional understanding that human beings, with body and soul, are "parts" of the society. Social systems are self-referential systems based on meaningful communication. They use communication to constitute and interconnect the events (actions) which build up the systems. In this sense, they are "autopoietic" systems.[4] They exist only by reproducing the events that serve as components of the system. They consist therefore of events, i.e., actions, which they themselves reproduce, and they exist only as long as this is possible. This, of course, presupposes a highly complex environment. The environment of social systems includes other social systems (the environment of a family includes, for example, other families, the political system, the economic system, the medical system, and so on). Therefore, communication between social systems is possible; and this means that social systems have to be observing systems, being able to use, for internal and external communication, a distinction between themselves and their environment, perceiving other systems within their environment.

Society is an exceptional case. It is the encompassing social system that includes all communications, reproduces all communications, and constitutes meaningful horizons for further communications. Society makes communication between other social systems possible. Society itself, however, cannot communicate. Since it includes all communication, it excludes external communication.[5] It has no external referent for communicative acts, and looking for partners would simply enlarge the societal system. This, of course, does not mean that society exists without relations to an environment, or without perceptions of environmental states or events; but input and output are not carried by communicative processes. The system is closed with respect to the meaningful content of communicative acts.[6] This content can be actualized only by circulation within the system. At the same time, but at another level of reality, the system uses the bodies and minds of human beings for interaction with its environment.

The logic of a theory of self-referential communicative systems requires this notion of an encompassing system as a limiting case. The theory of social systems, by its own logic, leads to a theory of society. We do not need political or economic, "civil" or "capitalistic" referents for a definition of the concept of society. This, of course, does not persuade us to neglect the importance of the modern nation-state or the capitalist economy. On the contrary, it provides us with an independent conceptual framework with which to evaluate these phenomena, their historical conditions, and their far-reaching consequences. In this way, we avoid prejudices toward particular facts; we avoid a *petitio principii*.

Types of Societal Systems

One consequence of this general approach is the way in which different historical types of societies can be distinguished. A society cannot be characterized by its most important part, be it a religious commitment, the political state, or a certain mode of economic production. Replacing all this, we define a specific type of societal system by its primary mode of internal differentiation.

Internal differentiation denotes the way in which a system builds subsystems, i.e., repeats the difference between system and (internal) environments within itself.[7] Forms of differentiation determine the degree of complexity a society can attain. Sociocultural evolution began with segmentary systems. Some of these societies developed a higher order of differentiation, above that of families or villages, namely stratification according to rank. All traditional societies that produced enough complexity to develop a high culture were stratified societies and, in this sense, hierarchical systems. Since these societies evolved from different regional sources, and since their aristocracies based themselves on land and/or cities, it was quite natural to conceive of different coexisting societies in spite of a certain degree of reciprocal awareness of each other's existence and of ensuing communication. The idea of society therefore assumed a territorial reference, however unclear its extension and frontiers.[8]

Modern society has realized a quite different pattern of system differentiation, using specific, *functions* as the focus for the differentiation of subsystems.[9] Starting from special conditions in medieval Europe, where there existed a relatively high degree of differentiation of religion, politics, and economy, European society has evolved into a functionally differentiated system. This means that function, not rank, is the dominant principle of system building. Modern society is differentiated into the political subsystem and its environment, the economic subsystem and its environment, the scientific subsystem and its environment, the educational subsystem and its environment, and so on. Each of these subsystems accentuates, for its own communicative processes, the primacy of its own function. All of the other subsystems belong to its environment and vice versa.

Basing itself on this form of functional differentiation, modern society has become a completely new type of system, building up an unprecedented degree of complexity. The boundaries of its subsystems can no longer be integrated by common territorial frontiers. Only the political subsystem continues to use such frontiers, because segmentation into "states" appears to be the best way to optimize its own function. But other subsystems like science or economy spread over the globe. It therefore has become impossible to limit society as a whole by territorial boundaries, and consequently it is no longer sensible to speak of "modern societies" in the plural. The only meaningful boundary is the boundary of communicative behavior, i.e., the difference between meaningful communication and other processes. Neither the different ways of reproducing capital nor the degrees of development in different countries provide convincing grounds for distinguishing different societies.[10]

The inclusion of all communicative behavior into one societal system is the unavoidable consequence of functional differentiation. Using this form of differentiation, society becomes a global system. For structural reasons there is no other choice. Taking the concept of the world in its phenomenological sense, all societies have been world societies. All societies necessarily communicate within the horizon of everything about which they can communicate. The total of all the implied meanings constitutes their world. Under modern conditions, however, and as a consequence of functional differentiation, only one societal system can exist. Its communicative network spreads over the globe. It includes all human (i.e., meaningful) communication. Modern society is, therefore, a world society in a double sense. It provides one world for one system; and it integrates all world horizons as horizons of one communicative system. The phenomenological and the structural meanings converge. A plurality of possible worlds has become inconceivable. The worldwide communicative system constitutes one world that includes all possibilities.[11]

In defining my concept of society, I carefully avoided any reference to social integration. The concept does not presuppose any kind of pooled identity or pooled self-esteem (like the nation-state). Modern society in particular is compatible with any degree of inequality of living conditions, as long as this does not interrupt communication. A self-referential system defines itself by the way in which it constitutes its elements and thereby maintains its boundaries. In systems theory, the *distinction* between system and environment replaces the traditional emphasis on the *identity* of guiding principles or values. Differences, not identities, provide the possibility of perceiving and processing information. The sharpness of the difference between system and environment may be more important than the degree of system integration (whatever this means), because morphogenetic processes use differences, not goals, values, or identities, to build up emergent structures.

Given its clear-cut boundaries, differentiating communicative behavior from noncommunicative facts and events, modern society is a social system to a higher degree than any of the traditional societies. It depends more on self-regulative processes than any previous society. And this may be one of the reasons why it cannot afford too high a degree of social integration.

Planning and Evolution

No society so far has been able to organize itself, that is, to choose its own structures and to use them as rules for admitting and dismissing members.[12] Therefore no society can be planned. This is not only to say that planning does not attain its goal, that it has unanticipated consequences, or that its costs will exceeds it usefulness. A first obstacle to planning relates to problems of observation and description. The observation of differentiated systems presents serious difficulties. Systems theorists normally presuppose hierarchical structures as a condition of in-depth observation and description.[13] Hierarchy, in this context, does not denote a chain of command, but the transitivity of subsystem building. Subsystems, according to this rule, are allowed to develop only within the boundaries of a subsystem. This expectation may, to some extent, be realistic at the level of organizations. It is highly unrealistic at the level of the whole society and its primary subsystems.[14] No pattern of differentiation, be it according to rank or according to function, can channel all further subsystem building into the primary scheme of differentiation. The society, therefore, lacks the inherent rationality required for its observation and, so much more, for planned change.

Planning society is also impossible because the elaboration and implementation of plans always have to operate as processes within the societal system. Trying to plan the society would create a state in which planning and other forms of behavior exist side by side and mutually influence each other. Planners have to use a description of the system, and will thus introduce a simplified version of the complexity of the system into the system. But this will only produce a hypercomplex system that contains within itself a description of its own complexity. The system then will stimulate reactions to the fact that it includes its own description and it will thereby falsify the description. Planners, then, will have to renew their plans, extending the description of the system to include hypercomplexity. They may try reflexive planning, taking into account reactions to their own activity. But, in fact, they can only write and rewrite the memories of the system, using simplistic devices which they necessarily invalidate by their own activity.

All of this, of course, does not prevent planners from being active and activities from being planned. By planning we are able to commit resources and activities in advance and to decide, more or less effectively, about the premises of further decisions. This may influence the state of the social system. We know how to handle production plans and electoral campaigns. We plan wars (defensive ones only, of course) and insurance schemes, school curricula, traffic flows, mass media programs, and many other things. Within small systems, and even within large organized social systems, chances are relatively high that activities are carried out as designed. This does not necessarily mean that effects turn out as intended. And it certainly does not mean that the society as a whole develops in a planned direction.

The societal system can change its own structures only by evolution. Evolution presupposes self-referential reproduction and changes the structural conditions of reproduction by differentiating mechanisms for variation, selection, and stabilization.[15] It feeds upon deviations from normal reproduction. Such deviations are in general accidental but in the case of social systems may be *intentionally* produced. Evolution, however, operates without a goal and without foresight. It may bring about systems of higher complexity; it may in the long run transform improbable events in probable ones,[16] and an observer may see this as "progress" (if his own self-referential procedures persuade him to do so). Only the theory of evolution can explain the structural transformations from segmentation to stratification, and from stratification to functional differentiation, which have led to present-day world society. And again: only observers may see this as progress.[17]

Whereas the post-Darwinian decades were fascinated by the alternative of creation (with author) versus evolution (without author), the idea of planned human evolution, in distinction to organic evolution, later replaced the first wave of social Darwinism.[18] Recent research, however, strongly suggests a third version of the relation between planning and evolution. Evolution itself can never be planned; this would be a *contradictio in adiecto*. But a self-referential system that tries to absorb planning may speed up its own evolution, because it becomes hypercomplex and will force itself to react to the ways in which it copes with its own complexity. If this is true, world society will have to face conditions in which more intentional planning will lead to more (and more rapid) unintentional evolution.

Paradoxes of Functional Differentiation

Problems are a consequence of the way in which a distinction between system and environment is made. Therefore, all of the most urgent problems of a societal system are the direct or indirect effects of its way of stimulating internal differentiation of systems and environments. In this sense they are, in our society, consequences of functional differentiation. They are the results of evolutionary developments, not the results of planning, and they are interconnected with all of the advantages of modern life. We cannot seriously want to change this condition of modern life; we cannot imagine an alternative to its mode of primary system differentiation; and in any case, we cannot plan to change the type of differentiation of our society.

We can, however, analyze the special risks we run with this type of society. Evolution is, as I have said, a transformation of improbable into probable states with increasing "costs." Without intending to "change the society," we can become aware of the relations between structures and their trains of consequential problems. Apparently, there are even self-defeating mechanisms at work. For example, functional differentiation both presupposes equality and creates inequality. It *presupposes equality* because it can discriminate only according to special functions (e.g., in schools according to school performance and prospects of further education) and because it operates best if everybody is included on the basis of equal opportunity in each functional subsystem (avoidance of exclusions, of *marginalidad*, and so on). But it *creates inequality*, because most functional subsystems (particularly the economic and the educational subsystem) tend to increase differences. Small differences in the beginning—be it in credit, in educational prospects, but also in scientific, artistic, and political "reputation"—become large differences in the end, because functional subsystems utilize differences and employ differences in pursuing their specific functions, and there no longer exists a superior mechanism such as stratification that controls and limits this process. The entire society, therefore, tends to proceed in the direction of increasing inequality; it accumulates differences between classes and between regions without being able to make use of these differences or provide functions for them, i.e., without being able to regress into the state of meaningful stratificatory differentiation.

Another example of this kind of built-in mechanism that may become self-defeating can be described as the relation between dissolution and recombination. Elements that formerly were regarded as natural units ("individua") have become decomposable, and their components have become available for recombination. We may think of the advances of physics, chemistry, and genetic biology, but also of the breaking up of persons ("individuals") into roles, actions, or motives as a consequence of advances in economic differentiation and organization. These advances, too, are consequences of functional differentiation. Dissolution or decomposition, however, not only provides *chances* for recombination, it also requires new forms of *control of interdependencies*. Singularized particles or motives (or even singularized persons) may associate in unpredictable ways. This problem has been underestimated; it was for a long time hidden behind distinctions of system and environment. To dissolve and to recombine were strategies of systems, and the changes of interdependencies came about in their environments. The famous problem of the "social cost" of economic production may illustrate this situation. Systems, generally, may control selected facts or events in their environment, related to their own inputs and outputs. They cannot control interdependencies in their environment. The more we rely on systems for improbable performances, the more we shall produce new and surprising problems, which will stimulate the growth of new systems, which will again interrupt interdependencies, create new problems, and require new systems.

It is a comfortable self-deception to attribute all of this to "capitalism." Capitalism in itself is nothing other than the differentiation of the economic system out of societal bonds, and it is by no means the only instance of functional differentiation. The concept of "capitalist society" makes it easy to locate structures in the system whose change would lead us toward a noncapitalist society, presumably a better society. From a systems point of view, however, this is a highly questionable procedure, because it is not possible to define the *unity* of a system by pointing to *specific structures within the system* which can then be changed. The unity of the system is the self-reference of the system, and its change will always require operating within, not against, "the system."

Evolution at the Level of Functional Subsystems

My argument can be summarized by two statements: (1) a functionally differentiated world system seems to undermine its own prerequisites; and (2) planning cannot replace evolution—on the contrary, it will make us more dependent on unplanned evolutionary developments. If this is indeed the case, then the prospects of further evolution deserve a second look.

There may be a continuing process of biological evolution on the level of human organisms, given society and culture as their environment. This is not my topic. Social systems are not a *late branch*, they are a *different level* of the evolution of order in general. If all social systems today belong to one single world society, the theory of evolution faces a new kind of problem: the level of sociocultural evolution is represented by one system only. There are no longer many societies from which evolution can select successful ones. A one-system evolution: is this possible? And is this possible without the almost certain prospect of destruction? In this situation, one alternative needs further consideration. Functional differentiation constitutes a kind of self-referential autonomy at the level of functional subsystems. This type of order, once attained, may set off evolutionary processes at the level of these functional subsystems. Within the general framework of the societal system we may have a plurality of evolutionary developments. The economic subsystem will evolve, but also the scientific subsystem, and possibly others too—each taking the others as the environment for its own evolution. The system of world society provides a sufficiently domesticated "internal environment" for its internal evolutions, whereas its own evolution becomes more or less dependent on the outcome of evolutionary processes within this internal environment.

In fact, if we scan the relevant literature, we find several attempts to reconstruct the history and development of functional domains in terms of concepts that are derived from a Darwinian theory of evolution.[19] Each subsystem may realize its own self-referential mode of reproduction—for example reproduction of a sufficient amount of liquid capital in the economy or reproduction of legal "cases" in the legal system—and may therefore find its own ways to deviate from its mode of reproduction, releasing processes of variation, selection, and restabilization. There may be different "accelerators" in different subsystems—for example, credit in the economic system, legislation in the legal system—increasing the chances for, and the speed of, structural transformations. This may "upgrade" the adaptive capacity" of the whole system,[20] but it by no means guarantees a viable relation between the system of society and its own natural and human environment. Evolution is unpredictable anyway. The separate but interrelated evolutions of the different functional domains within our differentiated society will reinforce this unpredictability. Their independence will bring about a higher degree of uncertainty with respect to the future. This makes it much more important than ever before to strengthen our ability to observe what is going on.

Self-Descriptions and Theories as Part of the System

Self-referential systems can, as the term implies, insert descriptions of themselves into themselves. These self-descriptions may be formulated at different levels of complexity: for example, these systems may "identify" themselves with a simplified image of themselves. Or, they may use a strategic difference to point to themselves, referring to one side and not to the other.[21] They may even conceive of themselves as "complex" and may orient themselves toward their own complexity—taking "complexity" as information about the lack of the information that would be required for a complete understanding and control.[22] They are unable, however, to objectify themselves and they will never be able to be available to themselves as objects. These remarks qualify the notion of *self-observation*.

Social systems, of course, are not self-conscious units like human individuals. Societies have no collective spirit that has access to itself by introspection. Self-observation on the level of social systems has to use social communication. Self-observing communication refers to the system that is produced and reproduced by the communication itself. In this sense, self-observation requires self-referential communication that refers both to the communicative system and to itself as part of the system.

To some extent, modern society had developed theories as instruments of self-observation within different functional sectors. During the eighteenth century, European society recognized new central problems of identity and order which arose in different functional subsystems. These problems could no longer be solved by stratification alone. This observation led to a new kind of theory, focusing on these identity problems, a new kind of theoretical reflection, differentiating itself along the line of functional differentiation. Political reflection had to take up the problem of sovereign power, its ability to decide all possible conflicts and its nevertheless nonarbitrary use. The solution was the constitutional state. The theory of cognition found itself facing the problem of the difference between subjective cognition and objective reality. Different solutions were offered by the common sense philosophers (Claude Buffier, Thomas Reid, David Hume) and by transcendentalism. The national and international economy required a theory of its own, focusing on production or on exchange or on distribution as basic models for the integration of economic activities. The theory of law had to recognize the fact that the whole of law is contingent on legal decisions and therefore on legal rules that regulate the production of legal rules. References to natural law had to be done away with and to be replaced by a "philosophy of positive law" (Feuerbach) or by purely historical foundations (Savigny). A theory of love became fashionable which saw love itself as responsible for its own troubles (and not parents, husbands, or other external circumstances) and focused on marriage as the solution. For education, the central problem was the increasing difference between human perfection and human usefulness, and it was solved, or at least alleviated, by a new concept of the individual.

There are fascinating parallels between these first waves of quasi-scientific self-observation within different functional subsystems. All of these theories were concerned with the reflexive foundations (e.g., basing law on law, education on education, love on love) and the self-referential autonomy of their respective subsystems. In this sense, they could claim universal, worldwide validity. Once differentiated, they had different motives for internal variation, for criticism, and for change. From Kant to Popper, from Adam Smith to Keynes, from Humboldt to Dilthey, from Feuerbach to Kelsen, from the theory of the constitutional state to the theory of the welfare state, more or less radical changes took place. But neither did these parallels become visible nor did these theories develop a self-referential framework to account for their own effects within their functional subsystems.

One interesting exception is love. By the eighteenth century, and perhaps even earlier, the semantics of passionate love reflected its own disturbing influence on real love relations. Reading about love prepares for love, stimulates doubts, creates inauthentic feelings and an awareness of one's own secondhand emotions.[23] The code of love reenters its own domain,[24] and its cultural imperatives become desperate, self-defeating rules which nevertheless have to be used to define relations as love relations.

Occasionally, we can find similar arguments in other fields. Savigny, for example, objects to the theory of positive law on the grounds that it will, if known and applied, undermine the confidence of the people, and will lead to rapid legal change and destroy legitimacy.[25] Substituting for the invisible hand the visible one of Keynesian planning may also become counterproductive. But such reflections on reentry are rare and tend either to destroy or to remystify order. On the whole, the pretension of "scientific" validity excluded the open admission of self-reference and circular reasoning.

Today, however, the theory of science itself is changing in the direction of a naturalized (neurophysiological, biological, cybernetic, sociological) epistemology which incorporates self-referential structures.[26] Universalistic scientific theories use concepts that also apply to science and to cognition itself—concepts like system, evolution, communication, complexity, meaning. The theories simply cannot avoid recognizing that they themselves appear within the world of objects that they describe. Despite many logical and methodological warnings, the recognition of self-referential systems is on its way.

These purely theoretical developments do not have immediate "practical" consequences. They may, however, change the ways in which the societal system can use theories as instruments of self-observation. The social structure and the semantics of modern society have grown in Europe. Their present shape is the outcome of evolutionary transformations using particular regional and historical circumstances. The impact of the European background remains strong, making all the more remarkable the fact that this tradition does not supply us with an adequate theory of society. For roughly one hundred years the materials have remained unchanged. The almost incredible revival of the theory of Karl Marx (1818–1883) confirms this thesis. Partial structures are used to characterize the whole system as a "capitalistic" or "industrial" or "postindustrial" society. Evolution is seen as an historical "process," although the theory of evolution treats only structural changes (and not processes!). Self-reference, on the other hand, is locked up in the "subject," leaving the "world" outside accessible for asymmetric technological exploitation. It is easy to see the interdependence of these semantic devices; they compensate for their differences on one side with overestimations and presumptions on others. They push society in the twin direction of technical and humanistic improvements; but they provide no theoretical framework for self-observation.

Systems theory has a certain capacity to improve the instruments of self-observation, i.e., of communicating within society about society. It is an international language, not designed to protect specific interests. Contrary to what is commonly thought of it, the focus of modern systems theory is not identity but difference, not control but autonomy, not static but dynamic stability, not planning but evolution. At least, there are remarkable advances that are changing the outlook of systems theory in this direction. However, these are developments within subsystems of subsytems of a subsystem of world society. It is difficult to see how they could become a common language for the process of societal self-observation.[27] Furthermore, systems theory, itself struggling to surmount the prevailing predispositions of the European tradition, is becoming more complex (and not simply more complicated in terms of models or variables). Evaluation and even understanding becomes difficult. Finally, there are no solutions for the most urgent problems, but only restatements without promising perspectives. Taking all of this into account, success seems to be highly improbable. On the other hand, we can see fascinating possibilities of arriving at a higher level of intelligibility. It requires, at present, a kind of stoic attitude to stay at the job and to "do the formulations"—*nec spe nec metu*. It may remain unsuccessful, but I cannot find it ridiculous.

Endnotes

1. See Aristotle, *Politica* 1252a:5–6.
2. In spite of many attempts to do so—from Hegel to Treitschke, to Leo Strauss, to Hannah Arendt. See St. Tm. Holmes, "Aristippus in and out of Athens," *American Political Science Review*, (1979), 73:113–128.

3. "He pason kyriotáte kai pásas periéchousa tàs állas." Aristotle, *Politca* 1252a:5–6.
4. See Humberto R. Maturana and F. J. Varela, *Autopoiesis and Cognition: The Realization of the Living* (Dordrecht: Reidel, 1980); and F. J. Varela, *Principles of Biological Autonomy* (New York: North-Holland, 1979).
5. The possibility/impossibility of communicating with God symbolizes this condition. See also Th. Luckmann, "On the Boundaries of the Social World," in M. Natanson, ed., *Phenomenology and Social Reality: Essays in Memory of Alfred Schutz* (The Hague: Nijhoff, 1970), pp. 73–100. Luckmann elaborates on the concept of the desocialization of the universe but refuses to accept the corresponding notion of a boundary of the social world.
6. At this point we are leaving the presuppositions of the biological theory of autopoiesis, using a different notion of "closure" and "autonomy." Biologists have to start with a definition of *life* whereas sociologists may use at this place a definition of *meaning*.
7. Obviously, a system can in this way serve as an environment for different subsystems which may emerge either simultaneously or one after the other. In each case, the internal environment is the part of the system that remains outside the boundaries of the specific subsystem.
8. See O. Lattimore, *Studies in Frontier History* (The Hague and Paris: Mouton, 1962).
9. See Niklas Luhmann, *The Differentiation of Society* (New York: Columbia University Press, 1982). For corresponding semantic transformations see also Niklas Luhmann, *Gesellschaftsstruktur und Semantik*, 2 vols. (Frankfurt: Suhrkamp, 1980–1981).
10. It is, of course, not very helpful to avoid this insight by distinguishing the global system and different societal systems on earth. This only leads us back into the unsolved problems of defining societies in a way that can be related to territorial units. And even if this problem could be solved by some kind of criterion, it would be difficult to see how this criterion could be related to our ways of understanding the typical features of modernity.
11. For the corresponding semantic transformation of the concept *world* see A. Koyré, *From the Closed World to the Infinite Universe* (Baltimore: Johns Hopkins University Press, 1957); and I. Pape, *Von den "möglichen Welten" zur "Welt des Möglichen": Leibniz im modernen Verständnis*, Studia Leibnitiana, Suplementa 1: Akten des Internationalen Leibniz-Kongresses Hannover 1966 (Wiesbaden: Steiner, 1968), 1:266–287.
12. For this concept of organization see Niklas Luhmann, "A General Theory of Organized Social Systems," in G. Hofstede and M. S. Kassem, eds., *European Contributions to Organization Theory* (Assen and Amsterdam: Van Gorcum, 1976), pp. 96–113.
13. See Leeuwenberg, "Meaning of Perceptual Complexity," in D. E. Berlyne and K. B. Madsen (eds.), *Pleasure, Reward, Preference: Their Nature, Determinants, and Role in Behavior* New York 1973, Academic Press, pp. 99–114. See also, for the manufacturing of artificial systems, Herbert A. Simon, *The Sciences of the Artificial*. (Cambridge, Mass.: MIT Press, 1969).
14. A good case study, relating to this problem, is Gunther Teubner, *Organisationsdemokratie und Verbandsverfassung: Rechtsmodelle für politisch relevante Verbände* (Tübingen: Mohr, 1978).
15. See A. G. Keller, *Social Evolution: A Study of the Evolutionary Basis of the*

Science of Society, 2d ed. (New Haven, Conn.: Yale University Press, 1931); D. T. Campbell, "Variations and Selective Retention in Sociocultural Evolution," *General Systems* (1969), 14:69–85; and K. E. Weick, *The Social Psychology of Organizing* (Reading, Mass.: Addison-Wesley, 1969).

16. See Niklas Luhmann, "The Improbability of Communication," *International Social Science Journal* (1981), 23:122–132 (chapter 4 in this volume).

17. It is notable that the use of "progress" (in the singular) became fashionable only around 1800, when modern society became visible and induced self-observing processes. See R. Koselleck, "Fortschritt," in *Geschichtliche Grundbegriffe: Historisches Lexikon zur politisch-sozialen Sprache in Deutschland* (Stuttgart: Klett-Cotta, 1975) pp. 351–423 (384 ff.).

18. See Julian S. Huxley *Evolutionary Ethics* (London: Oxford University Press, 1947); and E. Jantsch, ed., *Design for Evolution* (New York: Braziller, 1975).

19. See, for the economy, A. A. Alchian, "Uncertainty, Evolution, and Economic Theory," *Journal of Political Economy* (1950), 58:211–221; J. Spengler, "Social Evolution and the Theory of Economic Development," in H. R. Barringer, G. I. Blanksten, and R. W. Mack, eds., *Social Change in Developing Areas: A Reinterpretation of Evolutionary Theory* (Cambridge, Mass.: Schenkman, 1965), pp. 243–272; H. Riese, "Schritte zu einer ökonomischen Theorie der Evolution," in B. Gahlen and A. E. Ott, eds., *Probleme der Wachstumstheorie* (Tübingen: Mohr-Siebeck, 1972), pp. 380–434; K. Boulding, "Toward the Development of a Cultural Economics," in L. Schneider and Ch. Bonjean, eds., *The Idea of Culture in the Social Sciences* (Cambridge: Cambridge University Press, 1973), pp. 47–64 (pp. 55ff.); and R. R. Nelson and S. G. Winter, "Toward an Evolutionary Theory of Economic Capabilities," *American Economic Review* (1973), 62:440–449. For science see P. Caws, "The Structure of Discovery," *Science* (1969), 166:1375–1380; D. C. Dennett, *Content and Consciousness* (London: Routledge and Kegan Paul, 1969); J. A. Blachowitz, "Systems Theory and Evolutionary Models of the Development of Science," *Philosophy of Science* (1971), 38:178–199; St. Toulmin, *Human Understanding* (Oxford: Clarendon 1972), vol. 1; D. T. Campbell, "Evolutionary Epistemology," in P. A. Schilpp, ed., *The Philosophy of Karl Popper* (La Salle, Ill.: Open Court, 1974), pp. 412–463. For the legal system see H. Cairns, *The Theory of Legal Science* (Chapel Hill, N.C.: University of North Carolina Press, 1941), pp. 29ff.; R. D. Schwartz and J. C. Miller, "Legal Evolution and Societal Complexity," *The American Journal of Sociology* (1964), 70:159–169; Niklas Luhmann, "Evolution des Rechts," in Niklas Luhmann, *Ausdifferenzierung des Rechts* (Frankfurt: Suhrkamp, 1981), pp. 35–52.

20. In the sense of the evolutionary variable "adaptive upgrading," which serves the A-function in Parsons' theoretical framework.

21. See the logic, based on distinctions and indications, of G. Spencer Brown, *Laws of Form* (London: Allen and Unwin, 1971).

22. For this notion of "complexity," see H. Atlan, *Entre le cristal et la fumée: Essai sur l'organisation du vivant* (Paris: Seuil, 1979), pp. 74ff.

23. See R. Girard, *Mensonge romantique et vérité romanesque* (Paris: Grasset 1961).

24. "Reentry" in the sense of Spencer Brown, *Laws of Form*.

25. F. C. von Savigny, *Vom Beruf unsrer Zeit für Gesetzgebung und Rechtswissenschaft* (Heidelberg: 1814; reprint Darmstadt: Wissenschaftliche Buchgesellschaft, 1959).

26. See W. S. McCulloch, *Embodiments of Mind* (Cambridge, Mass.: MIT Press, 1965); Campbell, "Evolutionary Epistemology"; H. von Foerster, *Observing Systems* (Seaside, Cal.: Intersystems, 1980); H. R. Maturana and F. J. Varela, *Autopoiesis and Cognition: The Realization of the Living* (Dordrecht: Reidel, 1980); K. Knorr, *The Manufacture of Knowledge* (Oxford: 1981); and Niklas Luhmann, "Die Ausdifferenzierung von Erkenntnisgewinn: Zur Genese von Wissenschaft," in N. Stehr and V. Meja, eds., *Wissenssoziologie* (Opladen: Westdeutscher Verlag, 1981), pp. 102–139.

27. See also Niklas Luhmann, *Politische Theorie im Wohlfahrtsstaat* (Munich: Olzog, 1981), for the analogous situation within the political system.

11.
The Work of Art and the Self-Reproduction of Art

The following analyses are guided by two abstractions.[1] On the one hand they ignore all differences between individual art forms. Whether literature or theater, plastic arts or music—all are relevant as long as social communication treats the object (by whatever criteria) as a work of art. My interest is the consequences of the differentiation of art according to the special code *beautiful/ugly*, and here the differences between the individual art forms are not immediately important.

The second abstraction concerns the perspective that governs the posing of the problem. It requires a more extensive presentation.

We can discern in reality certain kinds of systems, which following a suggestion of Humberto Maturana are called "autopoietic" systems. These systems produce the elements, of which they consist, by means of the elements of which they consist. It is thus a question of self-referential closed systems, or more exactly of systems that base their relation to their environment *(Umwelt)* on circular-closed operational connexions. This kind of self-reference involves not only reflection, i.e., that the system can observe and describe its own identity. Everything that functions in the system as unity receives its unity through the system itself, and this applies not only to structures and processes but also to the individual elements that for the system itself cannot be further broken down.

It is not difficult, following this theory, to define the society as an autopoietic system. It consists of communications that are made possible and reproduced by the communications of which it consists. What is regarded and treated as the unity of a communication cannot be pregiven by the environment but is given by the connexion with other communications—above all by the conditions of meaningful negation (reflection). The concept of autopoietic systems thus fits society as a whole and at the same time this concept leads to an unequivocal delimitation of the societal system in relation to its environment, in which there is no communication. The question is then whether this is the only case of autopoiesis in the sphere of social systems, or whether and under what social-historical conditions other social systems can also attain this structural form of self-referential closure and autonomy in the constitution of their elements.

My hypothesis is that the structure of modern society makes it possible to form autopoietic subsystems. The way in which this occurs is determined by the functional differentiation of the social system. It seems to be the case that not all functional systems have reached the degree of autonomous differentiation that allows autopoietic self-reproduction. As a logical, lawful compulsion is not apparent, we must therefore consider from case to case whether and at what level of development functional systems not only reach a certain autonomy and capacity for regulation but also produce the elements of which they consist.

This can be shown with sufficient clarity for the legal and the economic systems of modern society. In the one case the system becomes autonomous through the communication of normative legal expectations, which can only be validated by reference to other elements of the same system. In the other case the system consists of monetary payments, which presuppose and permit monetary payments. This cannot be fully explicated here. In any case it cannot be seen as chance that precisely these two functional spheres dispose over a highly developed systems technology and were able to represent society in the liberal phase of the development of modern society.

For all functional systems the same question can be asked about the connection between differential and self-referential closure as the basis for an open and complex relation to the environment. Only when this connection can be made can closure and openness increase and become more complex. The theme of the following considerations is directed to one of these functional systems, the social system of the production and reception of works of art (the system of art). And in this limited framework I shall only be able to treat some problems that arise when this system achieves autonomy over the determination of its elements, strives for self-referential closure and precisely thereby seeks to development its sensibility in relation to its environment. As against Adorno it is a question here not of "autonomy vis-a-vis society," but of *autonomy within society;* we see the social nature of art not in negativity, in an "oppositional position towards society,"[2] but in the fact that emancipation for a specific function is only possible within the society.

Correspondingly, the autonomy of art attained in modern society is not something that excludes social dependence, not something that drives art into a hopeless marginalization. On the contrary: art shares the fate of modern society precisely because it seeks to find its way as an autonomous system.

That art has become differentiated as an autopoietic functional system in modern society shows itself particularly clearly in the fate of all attempts to call into question the traditional criteria of the beautiful, functions of representation, even the symbolic quality of works of art. This calling into question becomes itself the execution of the autopoiesis of art. The denial of all expressive intentions is thus understood as a particularly refined and elusive expressive intention—despite all assertions to the contrary. The reduction to mere objects, if that is the intention, does not escape the "frame effect."[3] Like every operation of a self-referential system the execution of the artistic operation must involve preconditions, even if it is only the precondition of fitting into the system. Even an unlimited arbitrariness in the choice of form and theme would not be able to alter this. The operations create "inviolate levels" and they are nothing other than the reference to the execution of the autopoiesis of art; one can seek to avoid every definition of this precondition and allow it to change with the operations, but this only makes it clearer that it is a matter of autopoiesis. The alternative would be: to leave the system.

If art with all its forms is seen as a social system and one asks about the elements of which these system consist, then one is led to the individual works of art. We could therefore suppose: art consists of works of art and what a work of art is is determined by art. Circular definitions of this kind are nothing new; they were commonplace as constituents of theories of good taste in the first half of the eighteenth century. Our problem is first of all whether the work of art is really the elementary unit of the art system, which cannot be further broken down. Sociologically speaking this would be an anomaly. For society already consists of communications (not for instance: of texts), and communications are events, not objects; and the economy also does not consist of commodities or capital but of payments. If we follow this line, then we can consider the work of art, if need be, as a compact communication or as a program for innumerable communications about the work of art. Only thus does it become social reality.

Works of art ensure in other words a minimum of unity and interrelationship (e.g., the possibility of supplementation of context) of the communications about them. They concentrate their connection. Alter understands within certain limits what ego experiences when, to put it in an old-fashioned way, he enjoys a work of art, that is, appropriates it. Communication about it, although it is in no way a question of a simple fact, can be correspondingly abbreviated. Communication tolerates and hides at the same time a high degree of discrepancy in what the participants consciously register and work through. The work of art unifies their communication. It organizes their participation. It reduces, although this is a highly improbable state of affairs, the arbitrariness of the foreseeable response, it regulates expectations. To submit to this with insight was once called "taste."

Without reference to a corresponding object this order in communication could not come about. This is banal in the sense that one could not talk about potatoes if they did not exist. However, the work of art is separated from the world of useful or dangerous things. It seems to be made specifically in order to provoke communication. It is not a question of a sum of isolated pleasures to be attained but of a socially arrived at judgment which has no other meaning beyond itself. In art communication becomes—one could almost say, to use a problematic concept—its own purpose. At any rate it is pushed to improbable and yet agreed-upon lengths. One's own response is experienced as guided, so that even the most intricate and esoteric subterfuge offers the prospect of reproduction, i.e., the possibility of consensus. This is the reason why explicit communication to a great extent need not occur, indeed can even be felt as inappropriate. Whoever advances and grounds his judgment of art is already in danger of appearing as someone who does not speak (superfluously) about the work of art but about himself.

The dissolving capacity of sociological analysis goes beyond the unity (wholeness, harmony, perfection) of the work of art. However, by this very means, it grounds a new understanding of this unity. Unity does not reside in the degree of centralization of the problematic, nor in the interdependence of the details, and certainly not in the risk of failure or in the avoidance of mistakes. These are all viewpoints that are not to be neglected: guiding viewpoints of production, auxiliary viewpoints of admiration, and reference points for the explicated discourse. But the unity of the work of art lies finally in its function as a program of communication, where the program can be so self-evident that it requires no argumentation and conveys the certainty of already having been understood. That is precisely what the theory of good taste seems to have meant when in its analyses of artistic judgment it emphasized the speed of the formation of opinions, immediate certainty, intuition, and the avoidance of all intermediary questioning by the understanding.

Now that we know what they are good for let us concentrate further analysis on the works of art themselves. It must be here if anywhere that we look for the key to the autopoiesis of art.

The work of art is both condition and obstacle for the autopoiesis of art. Without works of art there would be no art and without the prospect of new works of art no social system of art (but only museums and their visitors)."New" means here, as it has since the seventeenth century, not only another example, but rather something that diverges from the foregoing and thus surprises. Genius lies in the accomplishment of discontinuity and it is clear that this temporal discontinuity presupposes a social discontinuity, i.e., the differentiation of art from the tutelage of other, above all religious and political interests.

In this conjunction of the new with the surprising and divergent more is involved than is immediately apparent. For whatever has to be new has for this reason no future. It cannot remain new. It can only be admired as that which was new. The social system of art is thus faced from this point on with the problem of the continual disappearance of newness. With this accords the view of art theory that holds that the work of art should be a self-contained, harmonic whole complete in itself, which guarantees its permanence in time through sovereign disregard for time itself. This still leaves the question, however, of what the individual work of art can then contribute to the self-reproduction of art.

That the individual objects are kept ready for admiration, repeatedly viewed, read, performed, and preserved as far as possible from destruction goes without saying. Their destruction or even their sale abroad would be an "irreplaceable loss." They are sanctified and secured with safety alarms. We cannot go on without them— but actually not with them either. Their prices are rising, their truth gains clarity, but our intercourse with them in the social system of art unexpectedly acquires another quality. Boredom creeps in and the official celebrations have almost the effect of a stubborn refusal of this state of affairs, the effect of a countermeasure or of a compensation.

This is not least a question of the formal qualities of the work of art itself. Form is unstated self-reference. The fact that it can, as it were, put self-reflection on ice shows that a problem has been solved. Form refers to the context that poses the problem and to itself at the same time.[4] It presents self-difference and self-identity together. Where this succeeds the impression of self-sufficiency is created. The work of art creates its own context. It seeks to harmonize form and context, to be the unity of this difference. The art form absorbs all reflections, and what it radiates back is only its own significance.

Further, the (aesthetic) form must be ambivalent to the degree that it gives pause and directs questions back to the work of art. It must stimulate the comprehension of self-reflection and thus also communication about the work of art. It has always been accepted and demanded that the work of art arouse "astonishment." The "aestheticization" of art requires in addition that only the work of art itself can answer the questions that it raises and that neither knowledge of its style nor of its function is sufficient as an answer. "Astonishment" is thus relieved of all kinds of functions of directing attention in the interest of religion, morality, and politics; it too is, so to speak, differentiated.

The particular qualities of the aesthetic form are functional for the organization of the experience of and communication about art. They are dysfunctional for the autopoiesis of the system of art itself. For how is it to continue? What does the individual self-contained work of art contribute to making other works of art possible? Where does the "organization" of autopoiesis lie if the work of art must put value on its own isolation? The egg produces a chicken in order to produce another egg. The work of art would be the chicken, but where is the egg? Or: what corresponds to the genetic material—always dependent on the environment and to an increased degree—that ensures the continuity of self-reproduction?

The question can be answered with the aid of the concept of style. I define this concept functionally, without reference at this stage to its use in art theory. The style of a work of art allows us to recognize what it owes to other works of art and what it means for other, new works of art. The function of style is to organize the contribution of the work of art to the autopoiesis of art, and in fact in a certain sense against the intention of the work of art, which aims for self-containment. Style corresponds to and contradicts the autonomy of the individual work of art. It respects it and despite this diverts surplus value. It leaves the uniqueness of the work of art untouched and yet establishes lines of connection to other works of art.

For this concept of style it is unimportant whether style is introduced as a means of observation, description, analysis, and criticism of works of art or whether it already codetermines their production, i.e., artistic "praxis." If this scholarly distinction has any relevance it does not apply here. The two levels have influenced each other at least since the early Renaissance.[5] At most one could say that the difference between observation and praxis is set up within the system of art and thus presupposes its differentiation, and perhaps also that something like style (or functional equivalents, e.g., rules and recipes) brings about a corresponding difference of levels, i.e., of operation and observation as schooled, experienced observation.

Our functional determination of the concept of style also avoids the much-discussed question: must styles dominate a whole epoch in order to fulfill their historical mission, or is this neither necessary nor desirable? This is more a problem for the writing of art history than for art itself. The problem of what a work of art says, assimilates, and influences beyond itself can be solved within the framework of a pluralism of styles, even at the limit in terms of "the personal style" of an artist. We don't have to carry the longing for unity so far that pluralism and eclecticism become pejorative concepts. On the contrary: art is perhaps better advised when it avoids the risk of a unified style and opts for multiplicity, as long as associations (eclecticism?) remain possible. The question of whether and under what conditions unified styles dominate whole epochs is therefore excluded. It can only be answered in any case if we have clarified what is to be understood by "style" and whether this phenomenon can actually support a special historical "compulsion to unity."

If the considerations of the last section apply, style cannot be simply decided by differences of form. Rather, it involves the work of art in its central statement: the manner in which form and context are related. It is the unity of this difference and the manner in which it is achieved that makes the style of a work of art possible. Here context is everything that constitutes the horizon of the work of art and regulates its references. That can and will also include negative references by means of omission, abbreviation, and abstraction. Also, the quoting of other works of art (often ironically; think of Stravinsky) or quoting that operates between art forms (think of the written quotations in Hann Trier's paintings) belongs only to the context. Style is not to be found in the quotation but only in the way it contributes as quotation to the form of the work of art (and not only as an element of the form).

Style can arise from the model character of individual works of art. It is thus possible for it to exist in an effortless and unreflected way. The church tower of St. Paul de Leon becomes the model for other church towers in Brittany. This is only possible, however, if copying is permitted, if the uniqueness of the work of art is not a condition of quality and if prescriptive production is not harmful. Etymologically speaking *copia* originally meant a positive evaluation. In rhetoric, for example, it expresses the abundance of forms and figures of speech, and only when they are readily available in printed form does the meaning of "copying" become negative. As long as copying is praised as drawing on the rich knowledge of perfect forms, the level of meaning of style lies in the similarity of works of art. This level is not clearly differentiated from the form and execution of the works. However, the reproduction process must substitute more abstract symbols for the original object and its context, and it is this that compels the reduction to characteristics of a "style."

The concept of style can already be applied etymologically to such processes. It refers, then—and this is the case for the official terminology until far into the eighteenth century—to the manner *(maniera)* of the work of art. A further clarification did not appear to be necessary. In any case the literature on problems of style since antiquity is concerned more with distinctions than with what the distinctions have in common, which would have been the self-understanding of rhetoric. Even when the prescriptive productions of works of art was reflected and value was placed on originality (but not singularity) this concept of style was still retained. It seems to have had, as the correlative of the rejection of art governed by rules, just as indispensable a function, but now as the compensatory concept to the rejection of rules, as it were.

On the one hand style is thus not a recipe or a program of decisions by means of which the correctness of execution and the correctness of judgment of a work of art could be assessed. This task was taken over historically to an even greater degree by the work of art itself. On the other hand not everything is acceptable. The proclamation of the sovereignty of the work of art raises the problem, like all proclamations of sovereignty, of arbitrariness and its control; and here there appears to be an increased recourse to nature and also to style at the beginning of the eighteenth century. The differentiation of the program of decision and of style was basically already decided when the pluralism of styles and the eclecticism of execution were first registered—that is, in the sixteenth century. The rejection of art produced according to the rules is only a continuation of the already observable uncertainty (contingency). The work of art then serves the mastering of contingency and its return to necessity,[6] and thus fulfills a function that could not yet be realized on the level of style. It is only consistent, then, that each work of art is accorded the right to be its own program. However: what now guarantees that there will be art at all, and that there will go on being art?

This problem was not at all acute at first because the continuation of art was bound into the social structure and thereby guaranteed. The higher classes and the already differentiated functional centers of religion and politics saw to commissions. They needed art to illustrate their importance, and this all the more as the Old World was already reaching its limits. In the semantics devoted to art we nevertheless already find "preadapted advances," adaptations to something that does not yet exist; and it is not by chance that this anticipation of the future can be traced particularly clearly in the discussion of style.

If it can be generally said that works of art produce astonishment and admiration, now the *kind* of astonishment intended changes. The miraculous, pompous, exaggerated style is replaced by the call for the simple, the natural, and the sublime. The concept of style includes both: it is still defined as manner or as a way of arousing interest, but the astonished interest, the pleasure, the agreeable feeling that style should arouse has now become autonomous. The work of art is no longer employed for the support and magnification of hierarchically superior meanings; it is no longer just decoration for churches and palaces. It no longer aims for that amazement, of which Shaftesbury said that it is "of all other Passions the easiest rais'd in raw and unexperienc'd Mankind." On the contrary: it reckons with artistic connoisseurs. Now art produces its own public, and the only question now is who can participate.

When this development is first to be observed and what caused it needs more precise research. At all events it is clear that around 1700 artists were interested in a public aware of specific questions of art and aesthetically experienced, and that this is more important than positive or negative judgments in individual cases. Admiration is not sufficient; it must be knowledgeable admiration. That means: art no longer consists only of the performance role of the artists; it requires, for only thus can it become a social system, the differentiation of a public specific to this system, the differentiation of complementary roles. The differentiation of a social system for art results, in other words, in the differentiation of the difference between professionals and the public.

At the same time the model of rhetoric disappears for the artist. Rhetoric had never demanded that the speaker share the attitudes and feelings that he seeks to produce in his audience. Precisely the bridging of this difference was his creative role, and it is this that is increasingly called into question toward the end of the seventeenth century—in love as in art. An authentic relationship is now called for and on this basis a community of views and attitudes, of enthusiasm differentiated as something special. Exclusion and inclusion need now needs to be freshly regulated without recourse to rules and prescriptions which are used and controlled independent of attitudes.

In other respects too this retreat of the rhetoric tradition, directed to a nonreading public, can be observed. For instance, redundancy can be arranged in a freer and more individual fashion. The addressee can no longer rely on formulaic stereotypes; especially in poetry and literature one can develop one's understanding through the text itself, and traditional formulas seem boring or even an imposition that underestimates the abilities of the observer or reader.

The system of art is thus placed under increasing, individually varied claims and needs new titles for artists. That they are artists now indicates only their self-allotted position in this functional system. They had to be, for instance, geniuses in order to distinguish themselves in this already differentiated sphere. For the same reason the manner in which art draws attention to itself changes. The old "amazement" has to be hotted up. And the question of aesthetic norms, which guide production and critical judgment, is confronted with new demands. Finally, the understanding of style is related to the self-regulation of this relation between professionals and the public and becomes temporalized in the course of this development.

Only when there is no style to be found must new kinds of solutions be found for the relation of artist and public and especially for the old problem of amazement—for instance in the form of provocative or plaintive art or in the form of the grim effort to laugh at people, who do not take it seriously or do not even notice that they are being laughed at.

The functional definition of the concept of style is at the same time an historical concept. By "historical concept" I mean that the concept is codetermined by an historical difference which it itself brings about. This does not exclude a functional definition but on the contrary presupposes it. The problem called for by the distinction between work of art and style is itself an historical problem. It is given by the differentiation of a system of art. And only in relation to this problem does the recognition and change of style (as opposed to the work of art itself) make a difference. Or more concretely: it is not simply the model character of the work of art itself that fulfills the function of style; rather, style becomes differentiated as a special level of intercourse with works of arts. And only by this means is it possible to individualize the perfection of the work of art (freeing it of its model character) and at the same time posit style itself as authoritative and changeable.

The problem of the differentiation of autopoietic subsystems arises only in history and only relatively late—certainly long after the existence of art. Long before the problem became relevant the concept of style was already ready for it. With it a difference of levels enters into the system of art. With it contingency as the possibility of choice of style could be formulated without making the individual work of art arbitrary: the work of art could assert its own necessity under the rule of a style (or even through eclecticism of styles). All of this does not, however, explain *how* style fulfills its function, let alone whether and what restrictions of possible styles can be derived from the function.

The key to the further consideration of this question is offered by the *temporalization* and *historicization* of unified styles, their reshaping into epochal concepts.[7] This turning point is usually identified historically: along with many other historicizations it carried through in the second half of the eighteenth century after Winckelmann had successfully used the concept of style as a means of art historical research.[8] But what had exactly changed? The phenomenon of epochal styles and their sequence had not been unknown earlier. It was for instance widely recognized that the burlesque style of a Cervantes had been directed against the chivalric romances (which appear in the novel only as reading matter).

But this polemical formation of style was not persuasive. It contradicted all of the idealizing praises used to recommend the work of art. It was also impossible to imagine how a style, directed at the destruction of an earlier style, could gain permanent validity; for what was the point of perpetuating the destruction once it had done its work? It was the old theme of *varietas temporum;* but one saw in it only a lack of permanence and perfection in the world which also affected art. Changes in views of art and of styles were registered, but they were seen in the light of differences of quality; decadence was noted or, conversely, past styles were given terms of abuse (gothic, baroque, etc.), which only later became the familiar style descriptions. Time and the change of styles is negatively accounted for in one or the other direction.

The historicization of styles in the second half of the eighteenth century finally breaks with the traditional conceptions of time which had always allowed the unity of the beautiful, the true, and the good to be thought of as the acme of perfection. Only now can the work of art fully lay claim to its own singularity; for the individual uniqueness of the work of art is the surest guarantee that art always produces something new. Only now a theoretical aesthetic begins to work with problems specific to a sphere—that is, to react through reflection to the differentiation of the functional system of art. The now temporalized concept of style is no longer suitable as a substitute for rules; and it no longer grounds in compensatory fashion the artistic value of the work of art. Instead it makes possible a consciously historically situated politics of style. Previously the politics of style—e.g., the Academie Française in Colbert's time—had been a politics of purification, concern for the selection and legitimation of those forms that guaranteed quality. Now it is a question of decisions by means of which one gains historical distance and defines what corresponds to one's own time. This is no longer meant prescriptively but assures only the framework within which the work of art acquires meaning beyond itself.

One of the most important consequences of historicization is that the *comparability* of works of art acquires a *temporal direction* and is thereby *limited*. Traditional art theory had measured *all* works of art against common ideals of perfection and then, as can be read in the lives of Vasari, placed them in temporal sequence. Time led to perfection—or as a time of decadence, away from it. Now historical placing takes priority and is located deep in the work of art, which compares itself with previous art, seeks and gains distance, aims for difference, excludes already given possibilities. By this means it defines its style or the style it relates to. However, comparisons directed to the future make no sense. The work of art cannot define itself by its distance from future possibilities which are not yet visible. It cannot seek to exclude forms which have not yet been conceived. Giotto could not paint "not yet like Raphael." And the historian who makes such comparison fails his object, history. Comparisons are possible for art theory but only when tied to the precondition of a classical ideal of perfection.

This is the case not least because art, through its disposition over styles, has the opportunity, which practically no other functional system (least of all religion) has, of breaking abruptly with the past. Because works of art are complete, art can consciously and ruthlessly create discontinuity. It is not compelled to continuity. It does not have to wait until the investments have been written off. It does not even owe its patrons continuity. It can immediately fulfill the wish for the new. This is why art often produces anticipatory signals in social evolution which can be read retrospectively as prognoses. This is all the more easy if changes of style no longer indicate differences of quality and leave the validity of works of art of the past untouched, where every work of art strives to be as good as possible. Seen from this perspective the work of art becomes an element in a development of style, in which there is an awareness of style as a completeable and replaceable unity. The work of art could contribute nothing to the development of style if this were not the case. It could only have an insignificant effect if style were an endless, infinitely continuable quantity. The temporal structure of style itself, its microtime, allows it to further construction or defy decadence, to operate avant-gardistically or nostalgically, and to engage the whole quality of the individual work of art for such a politics of style. At the same time the work of art is and remains independent of style; and yet it can be good. And it is precisely the value of the works of art of a style which gives the latter its historical weight.

Style is thus, we may say, what joins work of art to work of art and thereby makes the autopoiesis of art possible. Autopoiesis, however, only means that *in general* art continues to be possible; it does not say *how*. Also, from a formal point of view, autopoiesis is possible in many different ways—we have seen, for instance, that a work of art becomes a model for many others. What leads then to closer determination of the structure, to more exact indications of the characteristics that a work of art "passes on" to others? What constraints are accepted in order to determine in which direction autopoiesis should go? It would be premature to answer this question by pointing to the pluralism of styles and interpret it as "anything goes." Certainly plurality and choice of style (both that which needs grounding and that which is capable of ground) has been important in this context at least since the sixteenth century. Concentration on just one style has been avoided (and probably at all times)—which does not exclude the possibility that such a unified style for this very reason has been sought for and considered desirable. But there are also other limitations that go beyond the pluralism of styles. They consist of *reactions of the system of art to its own differentiation and autonomy*, i.e., of a taking into account, within the system, of the fact that the system of art must provide for its own autopoiesis not in a free and arbitrary way but in a social environment.

In this sense style functions as the level of contact between the system of art and its social environment. Here the system of art must define, limit, and defend the closed nature of its reproduction and the autonomy of its choice of structure. Here it must refuse the claims of "interested parties" and in just this manner react to society. Here it must make evident its own work logic so that it becomes clear why art cannot be made to measure or ordered simply according to taste. The decisive insight is this: that this does not create arbitrariness, it simply defends "artistic freedom," but that this need to define autonomy *determines* important characteristics of styles, that is, *removes them from arbitrariness*. Every theory that simply opposes freedom to compulsion is a failure, given the complexity of the situation. From the perspective of systems theory the autonomy of reproduction appears rather as a burden or in any case as a constraint to use the *difference* of system and environment for *self-determination*.

As long as style is considered as rules or prescriptions for the correct production of works of art this level of the fixing of difference to society cannot unfold. It is assumed that society will enjoy beauty directly and that it is only a question of mastering the difficulties involved in the production of works of art. This changes in the eighteenth century, and the new historical consciousness of style demands that the given social situation of art be reflected. Romanticism is perhaps the first fully marked example. It celebrates "infinite reflection" in the work of art itself. It uses as context settings in which it knows that nobody in the social environment believes. It celebrates paradoxes. It cultivates irony. The inclination to quote art in art begins (as distinguished from the reuse of historical elements of style).

The subsequent realist and naturalist movements of style also face the necessity of pursuing a politics of distance. Precisely if one follows the stylistic program of confronting reality then it must be made clear in the work of art that this alone is not yet art; the comprehension of reality must also be made apparent and it must be shown that the work of art is finally due to the employment of specifically artistic means.

The retrospective definition of historically superseded art forms as "styles" also serves the same function. It too suggests, at least for the nineteenth century, a kind of stylistic consistency, which is increasingly used in a puristic manner, faithful to the original. What had a certain justification for the completion of the Cologne cathedral or the rebuilding of Carcassonne is generalized via the concept of style to the extent that it can also apply to new buildings. Then, in architecture at least, style can be offered by catalogue; but if you order it then you must follow the logic of the choice of style down to the details.

This kind of attention to style elements in the production and judgment of works of art still has the characteristics of a program that guides the selection of action and experience. It has not only "ideological" but also concrete relevance. However, it is not a program of decision in the sense that the quality of the work of art is simply given by the execution of the program. The individualization of the work of art demands rather that the work of art be its own program in a concretely deciding sense, i.e., that it itself delimits what is possible for it. Style is then, just as much as material, the limitation of self-selection—and in both cases it is a limitation that requires ostentatious emphasis.

This double function of style—on the one hand the ensuring the production of the elements by the elements of the same system and on the other the delimitation of the field in which this occurs—exactly corresponds conceptually to the definition of an autopoietic system. If this outline of the argument were to be developed and empirically confirmed, it would show that art too differentiates itself as an autopoietic subsystem in the course of the evolution of modern society; or at least that it is obliged to attempt this because it cannot otherwise maintain itself. From this hypothesis follows a corollary which concerns the "inclusion" of the public in the system of art.[9]

Above I indicated that art theory, as regards the social context of art, occupied itself at first with the general function of arousing amazement. The recipient must first of all be startled and astonished so that he is stimulated to experience and respond to art. This aspect is refined in a development of theory in the first decades of the eighteenth century, which attains its final form by reference to a public-specific art. The capacity for judgment of this public, formulated in the concept of taste, is assumed as a natural capacity, but already seen as something that must be acquired through reading or conversation in the salons or simply through the exercise of critical judgment. This changes with the further differentiation of the system of art and with the understanding of "style" as a mediating level between art and society. Demands on the public are increased.

If works of art are also expected to realize styles despite their self-sufficient isolation, then the connoisseur must be first of all a connoisseur of styles. He must be able not only to distinguish style but also to judge the correctness of style in detail, that is, be able to recognize breaks in style. This does not mean that he must or should judge as a purist; but he must be able to judge whether breaks in style are justified by the individual intention of the work of art.

This is, however, only a superficial, possibly dispensable requirement. More importantly, the work of art itself—as a component of its style—must increase the demands placed on participation, and this to an ever growing degree since the nineteenth century. The visual demands that painting makes of the observer since the nineteenth century become ever more complex and cannot be naively fulfilled. They demand broken immediacy and thus a particular distance from the object. With literature one has to think of the emergence of romantic irony and, further, of intenser forms of confusion of the reader.

The decoding of the work of art in terms of what is art in it requires trained activity; and this is not only a regrettable side effect of the complexity of the work of art but the grounds of its quality; it is its inherent intention, the requisite of this continued effect, the explication of its style. Further stages are reached when the transition to another medium becomes required and demands a corresponding training. Thus modern music is only understandable with the aid of the score (who cannot hear must read), while conversely the highly refined rhythm of modern poetry is scarcely readable; one has to hear it performed (who cannot read must hear). As a result the increase in the demands on inclusion has the effect of exclusion.

All of this is of course conditioned by the demise of rule-governed aesthetics. If there were rules and if they were followed and if this were the style then nothing more would be required of the public than the understanding of the application of rules. But toward the end of the seventeenth century it was felt that this was not pleasurable enough and so art was given license to discover how it could please. But art freed itself from this, driven by the necessity of assuring itself of its autopoiesis and thereby its style.

A final judgment on this development is not possible yet. Its effects can be observed as a reduction of the communication system of art to a narrow circle of admirers and as a greater differentiation of the art forms, in the sense that an understanding of modern poetry does not mediate an understanding for modern painting or theater or in particular for art direction in the theater. It does not follow a priori, however, that this must remain so and above all that this situation must be determined by decisions of style. It is clear enough that this would imply significant prejudgments about the future of art.

Are there functional equivalents for style? Are there other possibilities of solving the same problem?

We could think of *fashion*. Fashion also constitutes itself within time limits and yet it too functions despite this. In fact around 1700 the question was already raised whether a judgment about art could be anything but a question of opinion and fashion. Is beauty subject to fashion, or is it something for all time? Diderot along with many others set himself this question at the beginning of his *Traité du beau*, and it reveals a certain indecision that he could not immediately answer it.[10] The problem becomes really acute when epochal styles last such a short time and succeed each other so fast that they have the effect of fashions. Are styles and fashions functional equivalent or even one and same phenomenon today?

One affinity between style and fashion obviously resides in the fact that fashion can also be generalized and can involve different areas. It is therefore necessary to undertake a more exact functional analysis of fashion in order to determine similarity and difference.

It is not easy to know what was meant when around 1600 people started to say "la mode" as well as "le mode." At first the transient and evanescent was certainly emphasized. This turned out to be omnipresent once it had been fixed by the concept. Not only clothes and manners but also linguistic expressions, religious feelings, the style of sermons, dietary habits, the refined manner of cutting meat, and even the preferred way of killing and being killed, the duel, revealed themselves as products of fashion. Everywhere people appeared to take pleasure in the transient and the new. And then it was soon seen that it was not only a question of pleasure but also of security, especially for extravagances or otherwise conspicuous deviations.[11] A lot can be dared and supported as fashion because it does not mean a long-term speculation. And by its success fashion turns the burden of proof around: whoever does *not* adopt it becomes conspicuous.

The effects of fashion appear when not only behavior but opinions about behavior become socially regulated. For this communication is necessary that profiles opinions, censures them, and varies them in relation to their distance from behavior. Anticipations and connections become possible; security can already be gained in a certain distance from the given if one follows fashion. To this extent fashion, if it is successful, cancels ridiculousness. It can normalize the strangest deviations, but not all at the same time.

If we see fashion primarily as a strategy of security for the unusual, a strategy that pays for risk of the unusual with the decline of the fashion, then it is understandable why people have hesitated to surrender all values (or even the most important) to it. It was not only necessary to rescue from fashion true virtue (note the necessity of affirmation); beauty too must not be surrendered to it. Below this level of values the situation is less clear-cut in relation to "style." One can insist that the actual function of fashion is not the same as that of style. Security as regards the unusual is not yet autopoiesis, is precisely not a guarantee for self-reproduction. And yet a peculiar community of function can be seen. Does not the development of style profit from fashion? Does not the possibility of betting on fashion serve the risks of innovation in the development of style? And above all: could it not be the case that the input of extravagance must be constantly increased in the succession of styles in order still to be able to offer formal innovations, so that style and fashion gradually converge? The autopoiesis of art would then have to accommodate itself to changes of fashion and the question would then be not so much what a work of art contributes to style but how one fashion in style provokes the next.

The temporalization of the complexity of the system, the transference of significance to succession, is also a reaction to the difficulty of nailing down beauty as a criterion for works of art. What functions is in fact only the operative code, the difference between acceptance and rejection as beautiful or not. On this level historical judgment that can easily be given precision creeps in as a substitute criterion. Styles and even works of art are incorporated in an historical spectacle and judged from this perspective. In one respect, however, change of style and change of fashion could still be distinguished: by the degree of toleration for the arbitrariness of the break. Styles seem to develop according to Cope's rule:[12] they start simply and modestly and end in confusing complexity. Recommencement is governed therefore by the law of simplification and by the need to regain clarity. This can be followed from the emergence of the classicist style to that of the functionalist style, i.e., over a period of more than 150 years. In this respect at least there were conditions of continuation in the succession of styles; the negation of the preceding did not give a blank check for the arbitrary. The question is whether this still applies, since the functional style prescribed simplicity. It is difficult to avoid the impression that what has followed is only possible by means of arbitrary modification, by the intensification of an individual principle, which can no longer be distinguished from a change of fashion.

Certainly the tempo of change has increased—so much so that change of style can no longer be explained by generational change. (It is rather the case that generations are to be determined and artists' fates explained by what style was the fashion in their youth. If this is already or will be the situation we should not let our judgment be guided by prejudice against fashion. Neither style nor fashion preclude the quality of a work of art. More attention should be paid, however—as it were, against the politics of style—to the question of whether the substantial work of art can fulfill its function of organizing communication through itself. A maxim of La Bruyere is opposite here: "A man of fashion does not last long because fashions pass; if by chance he is a man of merit he is not destroyed and he continues to exist somewhere; just as estimable as before, he is simply less esteemed."[13]

Of course this supposes that the difference between *in* and *out* does not completely supplant the difference between *beautiful* and *ugly*.

Style, as we have known it for two hundred years, appears to be dissolving into fashion; at the same time it is threatened by the danger of history. In order to perceive and correctly estimate this danger it is first necessary to comprehend the new function of history. More and more it serves merely as a *proof of the contingency* of what has established itself. It is being restored, cultivated, preserved, and protected against its ordained end at enormous cost as a mode of the self-doubt of the present. Old music is celebrated on old instruments, although—and because!—the development of instruments permits a better sound. Factories in the style of Tudor castles are rescued at least as facade; that is quite sufficient. Behind the facade in Bielefeld, for instance, is a supermarket. The last horrors of the *fin-de-siècle* become objects of vehement communal politics. Spinning wheels and steam locomotives, no-longer-usable pithead towers, wooden kitchen boards, and copper kitchen utensils—the past deluges the present in order to demonstrate that it doesn't have to be as it is. It is obvious that this has nothing to do with artistic interests or aesthetic qualities.

Protest is sufficient and here one must have enough understanding in order to be able to evaluate the pressure exerted in this way on aesthetic experience. The relics of past cultural efforts, whatever their immanent value, are being used as witness against the present. If they weren't technically perfect—so much the better. If they look somewhat helpless and decrepit—all the more convincing. If they line up with the other disadvantaged and underprivileged—that is exactly the intention. Besides, it is not a question any more of their original function, and thus they can be allotted a new function at odds with all past valuations and instructions for use: that of the proof of contingency. The view of the past changes according to what one wants to see in the present. If you see the present as progress, the witnesses of the past with their insufficiencies demonstrate the appropriate "not yet." If, on the other hand, you want to document the problematic state of society and how little future content it has to offer, then the past is drawn upon in order to show that things could be different.

The "museumizing" of art appears imperceptibly to be taking on this function. The simple and the clumsy, set in the right light, demonstrates against the overrefinement of a late age; religious painting now says: I don't want to paint after nature, why should I? Strikingly, it is more difficult to show this state of affairs with literature and music—perhaps because printing had made a greater simultaneous presence of the noncontemporary possible, and there was always a choice. All the same: the remarkable Hesse renaissance shows that the same motives are at work. Whatever the case, my argument is that the function of appealing to historicity inhibits purely aesthetic interest and makes unpresupposed seeing and hearing more difficult, if not impossible. This applies all the more as the organizational side of the art industry has taken up this historical interest and oriented its marketing strategies to it—with the result that art is always displayed with the implied expectation that it be experienced from the viewpoint that it would not be possible like that today. What is left over consolidates itself as the avant-garde.

In this way art collections turn into museums. If, additionally, change of style and change of fashion converge and the resistance to the "museumizing" of art practically disappears, then the reproduction of art appears to be hardly dependent any more on a broadly based stylistic program. Difference and estrangement are enough. This may have the practical advantage that museums can buy works of art in the expectation that they will shortly attract interest as historical objects. Being in fashion and being out of fashion fit nicely together—at any rate in the calculations of the organization.

The results obtained do not allow us a sure judgment on the situation of art in modern society. They leave the question largely open whether, and to what extent, the social system of art as an autonomous, self-reproducing autopoietic system can be differentiated. They merely indicate some problems that must be registered in such a development and that are already noticeable to a considerable degree. A social differentiation of fine art into a social system with its own functional autonomy is not just, if at all, a progress to be welcomed. There is also an intensification of the internal problems in the sphere traditionally cultivated as art. Society continues to pay subsidies but withdraws its guarantee of continuity from an ever-the-same fine art. The internal constellation of art—seen with regard to society and not in terms of the multiplicity and diversity of works of art—has become more complex. It accommodated itself to a dependence on time, and from this follows the temporalization of works of art themselves. Even —and especially—the old works are assigned an historical place. The superiority of Doric columns is discussed, a taste for the archaic developed. Translated into the language of sociology, we can advance the hypothesis of the relation of several variables, i.e., the relation of:

1. the increasing differentiation of a functional system for art;

2. the increasing autonomy and responsibility of this system of professionals and the public for the self-reproduction of their communication by means of works of art;

3. the development of self-reflective theories, of "aesthetics" in the new sense, for the control of this autonomy and for the solution of the specific (incomparable) problems of this sphere, thereby setting in motion

4. the individualization of works of art, until finally by the middle of the eighteenth century the uniqueness of the object becomes the condition of recognition;

5. as a result: the problematizing of self-reproduction and the application of the difference of levels of meaning, concerned with the distinction between style and work, to the solution of the problem;

6. the use of the level of meaning of style to ground autonomy of art in its difference to (and this means at the same time dependence upon) the social environment and to include-exclude participants; and finally, this produces:

7. the complete temporalization of the complexity of the system, with the result that

8. the time-bound value of styles and even of objects threatens to displace the objective criteria of beauty (which, if at all, could only be determined self-referentially).

Through all of this, art mainly makes difficulties for itself. It thereby also betrays uncertainty in relation to its social function, the main focus of system-immanent reflection in other functional systems. The usual response has been to consider (and reject) a service function for other functional systems—rightly, precisely since it can not be the function of a fully differentiated functional system to contribute to another functional sphere. If analyses based on systems theory and social theory are employed, one arrives at very different departure points for a functional analysis and for observing and describing the self-descriptions of the system of art. This may provoke a rethinking of the traditional premises of aesthetic reflection.

The point of departure is communication in relation to an unusual object, disconcerted communication, which is directed back and bound to the object. Does art therefore serve the testing of communication by means of a case specially created for the purpose? Does it attain by means of the convincing object the limiting case of communicative community, which can dispense with communication to such a degree that communication only disturbs it? Is this a luxury that one can occasionally afford or is it the extreme point from which everything else can be opened up to communication? Is it not precisely the fictionality of art that, beyond all definite statements, lends—as medium, as it were—to the world a touch of the unreal? And is it not precisely the stringency of the work of art that assigns everything else the character of the "not really necessary" (without having to talk about alternatives at all)?

I leave completely open the question of whether a self-reflective theory useful for art itself can be developed from such considerations and whether it could still be called "aesthetics." Given such a state of affairs it is understandable that any social function of art is frequently disputed and its autonomy equated with absence of function. You can sign the death sentence in this way. Or you can revise the foundations of theory.

A sociology that conceives of modern society as a functionally differentiated social system does not assert that all functions resulting from functional differentiation work equally well. It has its doubts about religion and it can also pose the question, in relation to art, of whether differentiation suits it and whether it can succeed in autopoietic self-reproduction. There is no answer to this question that can be derived from theory. Faced with self-referential states of affairs, methodologically given asymmetries of deduction and causality are inadequate. One can only proceed as I have done, that is by seeking to discover what difficulties result from such a development and which functional alternatives are still available shortly before all options are closed off.

Endnotes

1. I exclude other possibilities of a theoretically based analysis of art, for instance the conception of art as symbolically generalized medium of communication. See Niklas Luhmann, "Ist Kunst codierbar?" in Niklas Luhmann, *Soziologische Aufklärung* (Opladen: 1981), 3:245–266.

2. Theodor W. Adorno, *Ästhetische Theorie* (Frankfurt: 1970), pp. 334, 335.

3. Erving Goffman, *Frame Analysis: An Essay on the Organization of Experience* (New York: 1974).

4. The distinction form/context is thus much more illuminating than the distinction form/content. See Christopher Alexander, *Notes on the Synthesis of Form* (Cambridge, Mass.: 1964) and George Spencer Brown, *Laws of Form* (New York: 1972).

5. Ernst H. Gombrich, "Norm and Form," in *Studies in the Art of the Renaissance* (London: 1966).

6. On this important tradition leading to secularization see Hans Blumenberg, "Kontingenz," in Kurt Galling, ed., *Die Religion in Geschichte und Gegenwart* (Tübingen: 1959), vol. 3, column 1793f.

7. For more extended treatment see Niklas Luhmann, "Temporalisierung von Komplexität in *Gesellschaftsstruktur und Semantik* Vol 1, (Frankfurt: 1980), 235–300.

8. Winckelmann's *History of the Art of Antiquity* is a classical example of the conception of history as the movement from the simple to the complex; here the concept of style acquires the function of ordering historical differentiation.

9. For other examples see Niklas Luhmann and Karl Eberhard Schorr, *Reflexionsprobleme im Erziehungssystem* (Stuttgart: 1979); and Niklas Luhmann, *Politische Theorie im Wohlfahrtsstaat* (Munich: 1981).

10. Adam Smith also had difficulties in limiting the influence of fashion on the judgment of beauty. See his *Theory of Moral Sentiments*, 5:1. And for Baudelaire fashion was half the game; without it the outcome would be an indefinable abstract beauty. See his *"Le peintre de la vie moderne."*

11. See on the dialectic of social security and individual profile, Georg Simmel, *Philosophie der Mode* (Berlin: 1905).

12. E. D. Cope, *The Primary Factors of Organic Evolution* (Chicago: 1896).

13. Jean de La Bruyère, "Les caractères ou les môeurs de ce siècle," *Œuvres Complètes* (Paris: 1951), p. 392.

12.
The Medium of Art

Works of art are not just traces left by human activity in the observable world. Neither do they arise as mere relics of purposeful behavior like tools, houses, street noise, or radioactive emissions. They serve, to take a minimal limiting criterion, the communication of meaning. This requires a medium in or through which communication occurs.

The following considerations attempt to discover something about this medium. The medium of *art*—I consciously ignore the difference between the arts on the assumption that with the help of the question of medium something in common can be observed and described. To do this one must operate at a level of abstraction that permits applications from the sphere of human perception in general,[1] ranging to questions of special symbolically generalized media of communication and even to questions of organization.[2]

Media differ from other materialities in that they allow a very high degree of dissolution. The original concept of matter—as opposed to form—had precisely this meaning: that which is undetermined in itself and thus is receptive to and dependent on form. For an ontological metaphysics, which worked with these concepts, matter was accordingly the medium of reality; then, also, the medium of a reality continuum of being and consciousness; and, finally, in so far as the world was considered as a *congregatio corporum*, the medium of a rationality continuum which, for example, alone made perception possible.

In this scheme of matter and form a second aspect was early added: the aspect of self-referentiality, by means of which form was revalued as mind *(Geist)* while matter as unreflexive being was relegated to the other side of the distinction. This posed the problem of whether all form was to be attributed to mind, i.e., was to be thought of as a construct of the self-referential processes of mind, or whether matter could itself attain form or thingness and how this, if at all, could be recognized.

This problem already entangled Kant in insoluble difficulties and contradictions.[3] For this reason I abstain from conceptual borrowings from this tradition. I speak neither of matter nor mind but confine myself to the concepts of medium and form. If a common metaconcept is required, then one can speak of a substratum. What is important, however, is that both substrata differ only relatively, that neither of them excludes self-reference, and that their difference varies historically, i.e., through evolution.

In order to stress relativity and evolutionary capacity I shall characterize media by their higher degree of dissolubility together with the receptive capacity for fixations of shape *(Gestalt)*.[4] This means: media also consist of elements or of events in the time dimension, but these elements are only loosely connected. Relative to the requirements of form or thingness they can be regarded as actually independent of each other. Thus money is a medium because payments can occur in any size, because a payment does not depend on the meaning and purpose of another payment, and because the medium is extremely forgetful (because, in order to preserve the value of money, it is not necessary to remember what the payment was made for) and only the ability to pay decides whether a payment is possible. But equally—to take another example—air is only a medium because it is loosely connected in this way. It can transmit noises because it does not itself condense to noises. We only hear the clock ticking because the air does not tick.

Forms by contrast arise through the concentration of relations of dependence between elements, i.e., through selection from the possibilities offered by a medium. The loose connexion and easy separation of the elements of the medium explains why the medium is not perceived but only the form that coordinates the elements of the medium. We do not see the cause of light, the sun, we see things in the light. We do not read letters but with the help of the alphabet words; and if we want to read the alphabet itself we have to order it alphabetically. Attribution is directed by the coordination of the elements, whereas the medium itself is too diffuse to arouse attention. It holds its elements ready for coordination through form. Heider thus speaks of external determination.[5] In the realm of forms, and to this extent the distinction remains relative, there can be more or less strict connexions, i.e., a dimension that shares both high elasticity and rigidity. Freedom of maneuver and elasticity of adaptation are thus preserved. A clock ticks and moves its hands, a ball bounces or rolls in reaction to impacts and conditions of its environment. A household can spend its money on different (but always specific) needs and a theory holds itself sufficiently undetermined and capable of adaptation in the logically coded medium of truth, so that it is not destroyed or unrecognizably deformed by every collision with reality. Works of art, especially those that require "performance" or depend on the effect of illumination and distance, must not fix their medium invariantly. Despite all of these relativizations, the difference between medium and form remains decisive *as difference.* There is neither a medium without form nor a form without medium. It is always a question of a difference between mutual independence and mutual dependence of the elements; and since it is a question of a difference this means that a relation of dependence of a higher degree is involved.

Besides these differences in structures of dependence, of interdependences, of loose and strict connections, differences of magnitude also play a role. Media consist of very many elements, in fact so many that every perception and every operative combination must proceed selectively. Forms by contrast reduce magnitude to what they can order. No medium gives only a single form, for then it would be absorbed as medium and disappear. The combinatory possibilities of a medium can never be exhausted; and if restrictions occur it is because products of form mutually disturb each other—for example, one noise shuts out another or one enterprise takes the market away from another enterprise; not, however, because air or money run out.

In the relation of medium and form the more rigid form asserts itself because it is less flexible. The unconnected (or weakly connected) elements of the medium can offer it no resistance. They are dependent on external determination. On the other hand form can only shape itself if a medium is available and its elements are suitable. Moreover, a form asserts itself at its own risk. It may suit it or not when it appears; and it remains exposed to disintegration or may suit itself to evolution, if it can reproduce itself.

This assertion of more rigid forms over less rigid repeats itself within forms. Sand adapts itself to stone and not the other way around. This too indicates the relativity of the relation of medium and form. A bureaucratic organization can be seen as form but also as a medium receptive to the imprint of interests. By accepting this relativity we obtain a point of departure for theoretical questions regarding evolution; for only then can we ask: how did physical evolution lead to form-structures (light, air, etc.) that are suited as media for perceptions that overcome distance, so that corresponding organisms can evolve; or, for the sphere of sociocultural evolution, how do language, writing, alphabetized writing, and symbolically generalized communication media arise, which hold ready a not otherwise available potential for form-structures that can be used once social conditions make this possible?

In order to produce form art is obviously dependent on primary media, above all those of optics and acoustics. It must be able to presuppose light and air. Beyond that, however, how can we say that art itself is a medium, a medium of communication? And if art is itself a medium, what, then, is form? In other words: what can we say about the relation of medium and form in the case of art? I have arrived at my theme.

Of course we must start by assuming that there is already a medium to which form can apply. For the case of art I want to test the opposite thesis: that form first constitutes the medium in which it expresses itself. Form is then a "higher medium," a second-degree medium which is able to use the difference between medium and form itself in a medial fashion as a medium of communication.

Let me exemplify this thesis in the case of music. There are many sounds that we automatically attribute to a source. The clock ticks, the telephone rings. The attribution to objects that cause the noise serves to direct follow-up experience and action. This also functions in the case of music. We get annoyed at radio music in the neighbor's garden and grab the telephone in order to stop noise through noise. In addition to this, however, the form of the musical work creates its own "reservoir" of selection, a space of meaningful compositional possibilities, which the specific work uses in a way that is recognizable as selection and that does not restrict other compositions. (Or does it? Is the medium in short supply? This we shall have to test.)

Even if music creates, with the help of instruments, pleasant-sounding tones, in this medium, again in the first instance, any tone can follow any other or be combined with any other, unless the form of the musical work decides otherwise. Here too through particular arrangements a medium is again first of all created in which form can imprint itself; here too we have loose and strict connections. The differentiation of composition and performance leads in addition to a special medium of notation, which was at first used only as a technical help but then was discovered as a medium for graphic forms which optically restrict what is musically permitted.

Music only functions as communication for those who can follow this difference of medium and form and can communicate about it; only for those who can also hear the uncoupled space in which the music plays; only for those who also hear that through its tonality music makes many more sounds possible then could normally be expected, and this in relation to disciplining through form. Art establishes, in other words, its own rules of inclusion, which are served by the difference between medium and form as medium. Whereas we normally hear noises as difference to silence and are thereby made attentive, music presupposes this attention and compels it to the observation of a second difference: that between medium and form.

It is clear that we can also apply this analysis to visual art. It too organizes for itself a medium in the natural world in order to separate itself from the world's conspicous events and play with its own. Through art, new possibilities of the acoustical and optical world are discovered and made available, and the result is this: strategies of dissolution permit more possibilities of ordering the world than would otherwise appear.

Finally, we can also maintain the same for literary works of art. The primary medium here is the alphabet. The alphabet permits combinations that are linguistically possible. Through the medium of alphabetical writing language can extend its own function as medium, it can be optically stimulated to new combinations of which one would not be aware acoustically in speech.[6] This applies to every kind of written language but can be increased if written language is used in order to create art forms. The same rule repeats itself here: artistic expression imprints itself due to its bound form in the medium. Only through it do we actually see how weak and arbitrary normal speech and writing are by comparison. Here too we have a relation of looser and stricter connection which can be used at the same time to open up spaces of possibility in language that language does not of itself offer.

Those possibilities were first discovered on the basis of rhythmic bonding, i.e., in direct continuation of the necessities of oral cultures. The alphabet then makes the reflected difference between prose and poetry possible. This is followed by ever greater freedom in the choice and combination of words until this difference again diminishes—to the point of the elegance that can be won when the normal word is freed of all the vague extensions of everyday language and used again in its exact original sense. The literary work of art leads to the discovery of language and then, not by chance, to the scientific form of this discovery: to a linguistics that sets itself goals beyond merely controlling grammar.

The distinction between medium and form competes with the distinction between entropy and negentropy and replaces it. The distinction between entropy and negentropy is current in art theory. It is faced, however, by the problem (to which Prigogine's theory of dissipative structures responds in a different way) that it can only encompass final states or, alternatively, tendencies but not processes of transformation.[7] If we add the distinction between medium and form, then the *dimension* that leads from entropy (chaos) to negentropy (order) can be considered as a relation of increase which makes possible more of both order and disorder.[8] A look at the history of art shows that natural media (media of perception) are always presupposed but that art in the process of its development creates additional media of its own in order to make use of differences. This can best be clarified by an example. The theme of the madonna and child changes in a process that is later described as the transition from the romanesque to the gothic "style." The child moves from the center, its meaningful place, to the side, where it is more clearly visible in contrast to the madonna; it becomes the element of a difference. In order to hold the child in this position the madonna is forced into a bodily balancing movement which accentuates the difference. Her body, her dress, her expression can be presented as autonomous necessity and at the same time as reference to the other, to the child. With the choice of this form the human body becomes a medium, i.e., the relatively elastic realm of possibilities from which form selects a certain (and no other) possibility. Form creates its own artistic medium by using it for its expressive purposes. Here too the rule applies: the greater rigidity asserts itself over the greater flexibility —with the risk that this assertion fails, is criticized, can be done better, or finally is relegated to history and the museum as the peculiarity of a certain style.

Let us take another example from the representations of modern technology or its objects in art. Where the severity of the intervention in nature (Cezanne's railway cutting) is shown and contextualized in nature, then nature itself can become a medium by making apparent that technology is one (but only one) of its possibilities.

In contrast to everyday primary experience nature is dissolved into elements that can be differently combined, and for this reason it is exposed almost (but not completely) without resistance to the intervention of technology and also art. In its own way modern science has discovered nature as a medium for the intervention of theories: as a medium that is open to different (but not arbitrary) possibilities of synthesizing. Compared to its successful sister art is set on seeing and doing things differently. It is thus inclined (but not necessarily) to judge technology negatively—in contrast to the presumed positive judgment by science. Finally, since the nineteenth century we can observe tendencies to constitute, with the aid of art, a further medium: society. As we no longer regard society as creation or as nature but as, we might say, its own concoction or, if we doubt the possibilities of planning, the result of its own evolution, here too it is possible to discover a medium. Sociology can occupy itself with society if it is looking for a medium that is suited for the methodically controlled construction of theory and constitutes it as a scientific discipline. But art also displays tendencies to appropriate the high capacity for dissolution and recombination of social data and to imprint its representations with the force of its own rigidity. Of course there can be no unanimous judgment on success or failure; however, the specific difficulties that result from such an artistic program can be examined more closely than before. It is self-evident that society as a system of its own operations cannot be a medium (since it can only be actualized in a structurally complex, selectively combined form). The question is thus: how can society behind society actually be projected so that society's choice of form can become grimacingly visible; and how can this occur in the specific manner of art so that the selection convinces as form and does not merely live, as social criticism, as the result of a momentary boom in "alternatives"?

My examples suggest that we describe the evolution of art as the increase in the capacity for dissolution and recombination, as the development of ever new media-for-forms. This would naturally require careful investigation and can only be presented here as an hypothesis. If it is the case, then the use of society as a medium would be the logical conclusion of such a development, its *non plus ultra*. Since art as communication is itself the realization of society, it could then use itself as medium and collapse in a kind of logical short circuit. Activities in this direction, in which finally everything is permitted in art, are not difficult to observe on a programmatic level. Even here, however, there are effective limits which lie in the inescapable need to use media of perception: words scattered over the paper must still be legible or at least visibly illegible, and modern music may only transgress the limits of hearability to the extent that this transgression is still hearable.

What this means is that art must presuppose socially constituted expectations—such as: writing is readable, music hearable, i.e., distinguishable from noises; or, simply, that what we encounter in concert halls, literary productions, museums, etc., is art. Without presupposing such expectations, however they are used or abused, art could not reproduce itself; it would dissolve into everyday life and drain away. But is the guarantee of perceptibility as art sufficient for the continuation of art as a social system, for the self-reproduction of art, for the "autopoiesis" of art?

In order to investigate this question (I do not claim to be able to answer it) we must use the concept of medium in an additionally limited sense, i.e., as a *symbolically generalized medium of communication*. We arrive at this concept when we recall that the difference between medium and form functions in turn as a medium, i.e., as a medium that opens up possibilities of the combination of media and forms for forming through communication.

A social medium only comes about when participants can observe (or assume that they can observe) what other participants can observe. It is thus always a question of second-order observation, of observations of observations, and this allows for the possibility of detachment from the direct, concrete evidence of observation. When we observe a work of art we can assume that the artist intended something when he made the work; and we can see that others assume that the artist could have intended what they assumed; and that, in turn, can lead the artist to talk about his art.

Communication about works of art is of course only possible if there are works of art. It would be false, however, to conclude from this a "first this/then that" relation, for the opposite also applies: works of art only exist if and insofar as possibilities of communication about them can be reckoned with. Once it is set in motion we have an autopoietic system which feeds on the production of works of art. Talking and writing are not the only communication; there is also shared perception, if it is actualized in relation to an object. Here it is not simply the normal presupposition of daily life, that others also see what one sees oneself, but it is a meaning-imparting looking (or hearing) which communicates at least to others that the object deserves attention.

The constructive freedoms of such a social system lie in the fact that only communication must function and that everything else is placed in the second rank of a necessary condition. We can express this in reflection through themes like "beautiful appearance" or "amazement." There is one condition that is decisive and has been termed "symbolic generalization" since Parsons in a sense that goes far beyond art.[9] The concept of generalization characterizes the capacity of the medium to overcome the differentness of things, i.e., its receptive capacity for different things; and the concept of the symbolic characterizes the condition of unification necessary if action (for Parsons) or, in our context, communication, is to occur at all. Here too we have, if in a somewhat different theoretical language,[10] a relation of dissolution and recombination, of openness to many possibilities and specific selection, of readiness for and dependence on the imprint of form.

In order to communicate about art we must presuppose the difference between primary medium and form and be able to make this difference into a medium. We must be able to recognize and use as medium the freedoms that the artist creates for himself for the choice of form. Communication about art is only possible on this basis, for it must be able to presuppose that there is information to be gained, and that means: that it would also be possible otherwise. Communication about art that is conditioned and limited in this way can occupy itself first of all with the medium that forms use in order to bring about distinctions. My examples: the possibilities of movement of the human body that are fixed by form as position or movement (dance). A radically changed problem arises when art wants to use society, of which it is part, as a medium. Communication about art is now placed in the situation of a medium with which art itself plays. Society as such, but also communication about art, must now be attributed with a structure in which events cohere scarcely or not at all, occur en masse and almost by chance, and are exposed to the intervention of more rigid complexes. This can be done easily (too easily) if a negative projection of society is made, since negativity possesses precisely this quality of nonconnection. Society may then be presented as figures without contact, as a bizarre ensemble, as arbitrary and fleeting constellations, and the form only serves to present this conception of its medium (which of course presupposes that it can do this and thus succeeds as form). The consequence, however, must be that art, insofar as it is realized as communication, must accommodate itself to this conception; then it will be able to mock its observers, exhibitors, and purchasers, and finally itself.

Dissolution can become an end in itself; the medium no longer serves the form but the form the medium until we arrive at the paradox that form only wants to maintain that it is its own medium, that it is not interested in itself. What happens, however, when society no longer accepts this and neither produces nor uses for self-description that entropic state in which it is nothing but a medium? We can take note that art offers such a projection; but we must also recognize that this offer contradicts our own projection, since this offer presupposes at least that communication about it is possible in a determinable way. Art taken to such an extreme behaves for the observer paradoxically and it thereby takes the narrow path on which the attempt to dissolve paradoxes can become fruitful.

Since communication about art (art as a social system) is dependent on given objects, it does not produce abstract conceptions about the ensemble of its possibilities. Although it is placed within the space of contingency created by art, communication does not need a concept of the set of possibilities, i.e., a concept of untranscendable limits. It must be able to distinguish between art and nonart and it can do this in the same way that houses and gardens can be distinguished without having to employ as a criterion a conception of the totality of possibilities of being a house or garden. It is only necessary to determine (and only to the extent that communication leads to consent or dissent) whether an object makes communication about art possible. To do this requires that the "reservoir" of selection is also seen and communicated. However, it is neither necessary nor possible to determine this "reservoir" as sum or set. It is sufficient that it functions as medium. The abstraction of these considerations shows that we have already reached the level of third-order observation and description, i.e., that we are engaged in the formulation of a theory whose object is the observation of observations. As this theory does not enter art as a work of art it must presuppose the autopoiesis of art. It can give no "thought impulses," let alone recipes, for the production of works of art.[11] As far as art is concerned it remains sterile. It can only become fruitful in the context of the autopoiesis of theoretical activity.

It is only on this level that we can meaningfully pose the question whether art can postulate its media at will or whether the possibilities of creating media are limited. It is clear that media of perception cannot be created at will, that visibility and audibility set barriers. The question becomes interesting in relation to society. Insofar as it refers to society art likes to permit itself a partly negative, partly utopian treatment of this material; and precisely when it depicts social situations "realistically," the pure duplication of reality translates the object into the mode of the made, i.e., the mode in which it could be made other. Other—but how?

Theoretically we can of course answer that society can be understood as a gigantic realm of surplus possibilities of communication and action, from which any one—and if any one, why not art as well—can select what works.[12] If this selection can be attributed to agents, e.g., the bourgeoisie, the ruling strata, or the combination of party ideology and bureaucracy, the gesture of rejection is relatively easy. It depends, however, on what it rejects, and makes itself dependent on tolerance. Despite this it is demonstratively bought or used as an object of speculation or it slips in some way through the gaps in the censorship of the regime.[13] But how and where does it find its society? And is it sufficient for it to see society as a medium without connections, as a medium for its own form?

If, however, society is neither nature nor the work of agents; if what works is what it itself makes possible; if society is an autopoietic system of self-selection, which also gives a place to those who believe that they can influence what occurs; if it is accordingly not meaningful to focus on rejection because this only blocks access to the medium; and if art can only operate in society and can only create in society a fictional reality, which can be turned against society: what is then medium and what is then form?

There is neither a prognostic nor recipelike answer to this question. The answer that follows from the previous considerations suggests, however, that only form can determine what is a medium for it and that dissolution cannot go far beyond what is capable of regaining shape. In other words, art must make use of form if it wants to show how far it is possible to dissolve and recombine something, just as it can only gain form if it presupposes an unconnected medium. The difference between medium and form can be taken to improbable lengths—but only within the limits in which the communication of form still succeeds.

Endnotes

1. See Fritz Heider, "Thing and Medium," *Psychological Issues* (1959), 1(3):1–34.
2. See Karl Weick, *Der Prozeß des Organisierens* (Frankfurt: 1985), pp. 269ff.
3. See Immanuel Kant, "Refutation of Idealism," in the *Critique of Pure Reason*, B 274ff., added later, with the "transcendental aesthetics" (B 33ff.).
4. In contrast to Herbert Spencer's theory of evolution, what is involved here is not a sequence, a movement from diffusion (dissolution into unconnected parts) to concentration and integration, but an evolutionary increase in the interdependence of *both* possibilities: dissolution and recombination.
5. Heider, "Thing and Medium."
6. It is now generally recognized that the alphabet influenced language itself to a considerable degree, e.g., through new words invented for writing and through the need for clearer syntactical structures. See Eric A. Havelock, *The Literate Revolution in Greece and Its Cultural Consequences* (Princeton, N.J.: Princeton University Press, 1982).
7. As Rudolf Arnheim also observes in *Entropy and Art: An Essay on Order and Disorder* (Berkeley, Calif.: University of California Press, 1971), pp. 26ff.
8. Cybernetics has correspondingly developed from an order-from-noise principle to an order-from-order-and-disorder principle. See Heinz von Foerstter, *Observing Systems* (Seaside Cal.: Intersystems, 1980), pp. 2–23.
9. For Parsons' most recent formulations see "Social Structure and the Symbolic Media of Exchange," in Talcott Parsons, *Social Systems and the Evolution of Action Theory* (New York: 1977), pp. 204–228, and "A Paradigm of the Human Condition," in Talcott Parsons, *Action Theory and the Human Condition* (New York: 1978), pp. 352–433 (392ff.).
10. I do not know if Parsons was aware of the concept of medium used by Heider. No efforts at conceptual approximation are apparent in Parsons' writings. I must therefore undertake my own attempt.
11. As was the expectation in the older cybernetic aesthetic theory. See Herbert W. Franke, *Phänomen Kunst: Die naturwissenschaftlichen Grundlagen der Ästhetik* (Munich: 1967), p. 110.
12. Such an approach can easily be applied to simple societies. See Elisabeth Colson, "The Redundancy of Actors," in Fredrik Barth, ed., *Scale and Social Organisation* (Oslo: 1978), pp. 150–162.
13. In the postscript to a volume of poems by Holger Teschke, *Bäume am Hochufer* (Berlin: 1985), written by a representative of the GDR, we read: "We made all this effort: totally different property relations, a brand new social order, a completely new state. He [the author] . . . doesn't want the world as it is, doesn't want it as we do but better, rummages in the past in search of a future, doesn't want to believe anything and thinks he knows everything. What have we produced here?" And yet Teschke's work is approved and published.

13.
The Self-Reproduction of Law and Its Limits

The predominant conception of the legal system refers to its organized and/or professional activities. People who do not work in the system appear as "clients," and thus the main question becomes how the system serves its clients. Critique of the system is incited by the demand for better service. Apparently, this demand does not have much success, since the system seems to resist and to repel all attempts to improve the service. The bureaucratic and professional ways of handling issues have to be taken as facts, and given these facts the critique changes its goals and proposes delegalization, deformalization, deprofessionalization.[1] Again, the results are not very convincing: they tend to make things easier and more difficult at the same time, probably for a different set of persons. Left-wing and right-wing critics, having lost their ideologies, vacillate between apocalypse and intrigue. The next step may well be more desperate and more radical claims combined with preadaptive resignation.

In such a situation a reasonable strategy may be to reconsider the theoretical foundations. Theoretical choices can be characterized by the kind of difference that they propose as the core problem. To select the difference between professionals and laymen or bureaucrats and the public for defining the legal system is a highly questionable decision—understandable in terms of everyday life but not in terms of theoretical refinement. For, to be clients of the legal system, people have to operate within the system. They have to be aware of a legal problem, have to define their situation accordingly, and have to commit themselves to advance legal claims or at least to communicate about them. They participate in the legal system using its system reference to give meaning to their activities. And even the decision not to use the legal framework for handling affairs of everyday life is a decision *within the system*. The legal system is responsible also for thresholds and discouraging effects.[2]

The difference between professionals and clients, seen as a difference of roles, or motives, or activities, or expectations, is an *internal* structure of the legal system. The legal system includes all acts or failures to act which are selected by reference to its mode of operation. Strategies of "delegalization" are at best proposals to restructure the legal system. They may, for better or worse, change the way in which law is taken into account. They may discourage people from using the law because of costs, congestions, or delays, or because appealing to legal remedies is no longer fashionable. But they will probably (and hopefully) not bring about a state of affairs in which the law is no longer acknowledged as relevant, giving legal and illegal behavior the same chance.

As theory, and in its practical results as well, the paid-work paradigm of professionals and bureaucrats leaves much to be desired. Above all, it lacks any clear understanding of the specific function of the law. Using the general framework of the theory of society as a functionally differentiated social system, we can conceive of the legal system as one of its functional subsystems.[3] Such a system constitutes itself in view of its function. The function is a problem that has to be solved at the level of the societal system. A one-function/one-system arrangement requires complete autonomy of the system because no other system can replace it with respect to its function. Hence, *autonomy is not a desired goal but a fateful necessity*. Given the functional differentiation of society no subsystem can avoid autonomy. Notwithstanding all kinds of dependencies and independencies in relation to its social and its natural environments that system alone can reproduce the operations that fulfill its function. Whatever serves as unit in the system, including the unity of the system itself, has to be constituted by the system itself. All elementary units (e.g., legal acts) and the unity of the system as well are achieved by the reduction of complexity. They are performances of the system itself and are never given to it by nature or by other environmental conditions. Therefore, given the general regime of a functionally differentiated society, all law becomes positive law which, of course, is not necessarily statute law but can also be created by the courts and by contract.

In this sense, functional subsystems of society are always self-referential systems: They presuppose and reproduce themselves. They constitute their components by the arrangement of their components, and this "autopoietic" closure *is their unity*. This mode of existence implies self-organization and self-regulation, but it has to be realized not only at the level of the structure but also, and above all, at the level of the elements of the system.[4]

This general theoretical conception can be applied to the legal system. If such a system evolves within the context of functional differentiation all regulation must be self-regulation. There may be political control of legislation, but only the law can change the law. Only within the legal system can the change of legal norms be perceived as change of the law. This is not a question of power or influence, and this not to deny that the environment and particularly the political system has an impact on the legal system. But the legal system reproduces itself by legal events and only by legal events. Political events (e.g., elections) may be legal events at the same time, but with different connections, linkages, and exclusions for each system. Only legal events (e.g., legal decisions but also events like elections in so far as they are communicated as legal events) warrant the continuity of the law and only deviant reproduction, merging continuity and discontinuity, can change the law.

A simple fact never bestows the quality of being legal or illegal upon acts or conditions. It is always a norm that decides whether facts have legal relevance. After many centuries of doubts and discussions we are today used to admitting that neither natural nor religious nor moral conditions have this lawmaking potential, but only legal norms. The legal system is a *normatively closed system*.

It is at the same time a *cognitively open system*. Following recent developments in systems theory we see closure and openness no longer as contradictions but as reciprocal conditions. The openness of a system bases itself upon self-referential closure, and closed "autopoietic" reproduction refers to the environment. To paraphrase the famous definition of cybernetics by Ashby: the legal system is open to cognitive information but closed to normative control.[5]

Normative closure does not exclude cognitive openness. On the contrary, it requires the exchange of information between system and environment. The normative component of legal meanings provides for *concurring* self-reference.[6] Concurring self-reference *is not a rule* as we are used to thinking as successors of Kant. Therefore the normative quality of legal operations cannot be reduced to the enactment or the application of a rule. Rather, it is the necessary and continuous reformulation of the unity of the system. Having its reality *in actu* it binds the following operations to confirm and to reproduce the system. It needs for this very purpose limitation and guidance—but not determination!—of choice. In this sense, concurring self-reference uses the difference of system and environment to create information. this would never work if the system were a system of norms only and its environment the realm of cognitions. The legal system is not a normative system if that means a system the elements of which are norms. It is a system of legal operations using normative self-reference to reproduce itself and to select information. The legal system, basing itself on its normative self-reference, is an information-processing system, and it is able to adapt itself to changing environments if its cognitive structure is sufficiently generalized.

Normative closure requires *symmetrical* relations between the components of the system where one element supports the other and vice versa. *Cognitive openness,* on the other hand, requires *asymmetrical* relations between the system and its environment. The operations of the system are contingent on those of the environment and adapt to changing conditions. The impact of the system on its environment, for example compliance with rules, is again an asymmetrical relation in which the environment adapts to the system. Both contingencies have to remain separate to avoid circularity.

For this reason normative structures are highly vulnerable. They are sensitive to open defiance and unenforceability because doubts have a spillover effect and spread over the system. Cognitive structures, on the other hand, may be specified and remain relatively isolated. The concurring involvement of normative closure and cognitive openness in legal events and operations combined symmetrical and asymmetrical, general and specific commitments. The emergence of a normatively differentiated system does not lead to a state in which cognitive orientations are less important. They become more important—for that system!

Other systems use other ways and other semantic forms to distinguish and recombine openness and closure. The economic system, for example, operates openly with respect to needs, products, services, etc., and it is closed with respect to payments, using payments only to reproduce the possibility of further payments. Linking payments to the exchange of "real" goods interconnects closure and openness, self-reference and environmental references. General purpose money provides for closure and remains the same in all hands. Specifiable needs open the system toward its environment. Therefore, the operations of the system depend upon a continuous checking of one in terms of the other. This linkage is a prerequisite for the differentiation and self-regulation of the economic system.[7] The same holds true for the legal system, of course with different mechanisms to provide for self-regulation by closing off self-referential procedures. In this case, the reference to the normative framework of the law serves to establish circularity within the system: decisions are legally valid only on the basis of normative rules because normative rules are valid only when implemented by decisions.[8]

The validity of law cannot be founded on authority or will, as the legal positivism of the nineteenth century was. Nor does the "fait social" of human society grant validity. It is not the "existence" of the legal order that is the source of itself,[9] and it is not the hypothesis of a basic norm that constitutes (or simply postulates) the object of legal cognition.[10] Austin, Durkheim, and Kelsen offer competing attempts to avoid circularity and to found the validity of law on something else. However, *validity is circularity*—circularity, of course, in need of logical unfolding.

Finally, the legal system, for its own reproduction of legal events by legal events, needs a binary structure in terms of which all events can be described as not being their counterpart. The system uses the code of right and wrong to duplicate all meanings—the right events being not wrong, the wrong events being not right. By this very description, whatever happens and whatever can be done becomes contingent. It remains possible to select the right or to select the wrong but not without committing oneself to negating the opposite value. The right path, then, may become a bit too righteous and the wrong may be overburdened with consequences stemming from the fact that it was not right.

Such a binary schematization is neither a fact of nature nor a law given by divine logic. It is an achievement of evolution, an evolutionary universal. At the time of the Greek tragedies it could not be taken for granted.[11] On the other hand, it is not simply an analytic device, structuring the recognition of the law by discerning right and wrong. It links the reproduction of the law with the reproduction of the contingency of the law and it serves as *prereq-*

uisite of conditionality (see below). Based on this prerequisite, the legal system can be erected as a network of conditions preprogramming events (and particularly actions) to be either right or wrong.

This outline of a theory of self-referential legal systems uses the distinction of normative and cognitive orientations to distinguish and recombine closure and openness. Any further development of this line of thought depends upon the way in which we conceive of norms and cognitions.

It will not be very helpful to define norms by self-implication—referring to the meaning of "should" or to the justification of sanctions.[12] This would give us a self-referential concept but not a theory of self-referential objects. To avoid all kinds of conceptual circles I start with defining the difference between cognitive and normative orientations as the difference between learning and not learning.[13] The problem of choice between learning or not learning arises in the face of inconsistent experiences.[14] If an experience contradicts our expectation we can either accept that fact and change our expectation; or we can try to maintain our expectation and to treat the experience as deviant, as an unusual exception or as a wrong choice. Since this problem of changing or not changing expectations comes up regularly in everyday life and particularly with respect to human behavior, it may be required that one commits oneself before the event and declares one's intention to change or not to change the expectation in case of contradicting experiences. The symbolism of cognition and of norms, of "being" and of "should," of "existence" and of "axistence"[15] provides general semantic forms and recognizable patterns for such commitments before the event. Using these forms, we may bind ourselves to expect either cognitively or normatively and, accordingly, to change or not to change our expectations in the case of disappointing experiences. And if this is possible we may build expectations of expectations; we may normatively expect normative expectations or may be normatively expected to apply cognitive expectations and vice versa.

Expectations of expectations can be called *reflexive* expectations. Normal expectations never become reflexive. I do not expect to expect to be able to start my car or to get a response at the reception desk of the hotel when I ask whether accommodation is available. Only if the car does not start do I feel that it would be inappropriate to stick at the expectation that it should start. And only if I get the answer "no rooms available" and discover later that it was false do I feel that I have a right to get at least an honest and true answer. Reflexive expectations are evoked only if there is a point in making a choice between cognitive and normative expectations, i.e., reflexivity depends on forced choice, so to speak, on the level of primary expectations.

Looking more closely at the research about reflexive expectations would lead to significant refinements.[16] For legal theory it is more important to connect this field of research with systems theory and particularly with the theory of self-referential systems. Social systems in general use expectations as structures that control the process of reproduction of communications by communications.[17] Therefore, differentiation of social systems requires the specification of expectations that maintain the autopoietic process of reproduction. The legal system becomes differentiated by distinct standards of self-reference in everyday operations. It uses the normative quality of expectations, i.e., resistance against learning, to include operations in the system and to refer to further operations (e.g., execution) of the system. It can associate further meanings as additional conditions of legal validity and it may try to warrant some kind of conditional predictability. But these normative meanings work as concurring self-reference only, assuring the reproduction of legal events out of legal events. There is no need and not even the possibility of complete self-determination. The legal system does not determine the content of legal decisions—neither logically nor by some kind of crafty procedures of hermeneutic interpretation. It operates as a closed and at the same time as an open system, normatively referring to the maintenance of its own self-reproduction and cognitively referring to adaptive requirements with respect to its environment.

If all legal events are normatively bound to push on the process of autopoietic regeneration and are nevertheless cognitively prepared to learn from the environment, the system will have to face up to problems of compatibility of these divergent and perhaps even contradictory attitudes. Such combinatorial constraints may bring about limits to the growth and the complexity of the system. Since closure and openness can be combined this is not a hopeless contradiction and not a real impossibility. But we have to specify what kind of mechanisms extend the realm of feasible combinations. And my presumption is that the actual symptoms of overstrain in the legal system are generated by these mechanisms as a kind of immune response against environmental (and particularly political) pressures and are not primarily problems of enforcement or problems of insufficient legitimacy or justice.

Mechanisms that differentiate and recombine normative and cognitive orientations work on two different levels: the one general, the other specific. At the the general level the system uses the fundamental technique of conditioning.[18] Special events (actions, decisions) within the system are activated if, and only if, certain other events are realized and thus are conditioned on preprogrammed information. Conditionality provides the chance to differentiate and recombine norms and cognitions. The conditioning program itself can be stated as a norm that foresees deviant behavior and does not become invalidated by it. Applying the program, on the other hand, requires cognitive operations. It relies on the capacity to handle information and to learn whether certain facts are given. With this kind of "iffish" attitude long chains of events can be built, each step depending on previous others and all depending on the legal validity of themselves and of others.[19]

In this sense, conditional programs are the hard core of the legal system (Willke gives a rather different account).[20] All legal norms are conditional programs and if they are not formulated that way they can be translated into if/then relations. This makes it difficult to confer to future states the status of a condition of legal validity.[21] Legal rules may mention future states. The prospects of the child's welfare should guide the decision about which of the divorced parents should take care of the child. But this does not mean that the decision and all acts based on it will lose their legal validity if the future falsifies the prediction. The decision depends on present informed guesses about the future, and legal validity is used (or misused?) to absorb risks and uncertainties. Law does not specialize in fortune telling, and the legal validity of gambling has always been a subject of suspicion.[22]

This, however, is only half of the truth and only one way to relate normative and cognitive components of the legal system. Conditionality is the general and indispensable device, but there are also more subtle, subcutaneous ways to infuse cognitive controls into normative structures. Judges are supposed to have particular skills and contextual sensitivities in handling cases. They apply norms according to circumstances, and if necessary generate exceptions to confirm the rule. They try to do justice—and postpone the perishing of the world from case to case. Other learning processes take place at the dogmatic level of legal concepts.[23] (Broekmann and Heller argue against the possibility of change at this level.)[24] The conceptual framework of legal doctrine adapts to changing conditions and changing plausibilities and it may reflect and control its own change because concepts are not yet normatively binding decisions.[25]

The actual problems within this area are more or less problems of time and of speed. The unity of the legal system requires an integration of changes on both levels: court decisions and legal dogmatics. New conceptual developments or new dogmatic rules have to wait for stimulating cases and cases can be aggregated into types of problems only if the conceptual development is sufficiently advanced. All of this takes time—and under modern conditions apparently too much time. Sufficient speed can be achieved only by legislation and legislation will change the law again and again without leaving time for court traditions and for dogmatic refinements to settle down. Within the legal system the priority passes on to the legislature. This means, to some extent, a new primacy of cognitive over normative considerations. The law has to fit the society around it and we are lucky if it nevertheless remains able to fulfill its own social function.[26]

From ancient times we are familiar with a critique of law that assumes several forms.[27] Law is unjust or at least not quite in conformity with the idea of justice. This seems to be inevitable as long as we need property. Moreover, legal norms are never completely enforced. We know of hidden and even open deviance. Lack of justice and lack of compliance and enforcement have to be taken as normal in this world. In both respects, ideal and material, the legal system lacks perfection. We may add the famous *ius vigilantibus scriptum* or its modern equivalent: the differential access to law. And last though not least we are aware of many ways in which sophisticated legal forms are misused to bring about effects that were not intended by the legislator[28]—one of the most famous examples of tricky misuse being the emancipation.[29] All of this remains on the agenda. In addition, however, we have invented a new kind of discontent. We feel that the legal system suffers from overstrain: that we have too much legal regulation, too much of the good.

What does this mean? I know of no legal theory that explains how the good turns into the bad and at what point. Sociologists could, once more, call up Max Weber. His analysis of bureaucracy shows how rationality can become a nuisance. But this *paradigma* is a rather impressionistic kind of theory. If we try to transfer this insight from bureaucracy to the law in general (i.e., from organization to a societal subsystem) the conceptual construction will break down. It is not strong enough for transmission.

Undeniable symptoms of overstrain have suggested that alternatives to the law or at least new ways of delegalization (including deformalization, deprofessionalization, etc., etc.) should be looked for.[30] This may suggest practical innovations, but it is wrong as a principle. A functionally differentiated society cannot provide for alternatives to its functional subsystems. All functional equivalents are part of the functional subsystems because these are organized in view of their functions. It is not possible to inaugurate functional equivalents outside of the system because being an equivalent includes them in the system. Moreover there is no way to speak of "alternatives" except in terms of functional equivalents. Otherwise, an alternative would simply be something else which may or may not have an impact on the system. The political system cannot replace the economic system, the economic system cannot replace the educational system, the educational system cannot replace the legal system, the legal system cannot replace the political system, *because no functional subsystem is able to solve the core problems of another system*. Functions are points of view for comparison and substitution and, therefore, society, which bases its differentiation on function, builds self-substituting *and not other-substituting* subsystems. Hence, each proposal of an alternative has to specify the function in view. If it is the function *of* the law it cannot stimulate an alternative *to* the law. If the proposal relates to secondary functions—say, slowness of procedures, insensitivity to personal feelings, overcentralization—there may be remedies available *within the system*. There may be also remedies *outside of the system*, but using them implies either using the means of other systems (e.g., using money to establish more courts to speed up procedures; using political power to suppress conflicts that otherwise would come before the courts) or reducing the domain of law, or both.

In general, the extensive discussion of overstrain and similar problems lacks clear statements about problems, functions, and system references. The present German debates about *Verrechtlichung* have more or less political overtones. They are part of the unanimous critique of the welfare state (whether this be the last resort of capitalists and technocrats or an attempt to ruin the economy by socialism).[31] The main suggestion is to look for self-healing forces of the society, to engage the lifeworld of small communities, the self-organization of discussion groups, and the reasoned elaboration of everyday activities.[32] However, this communal approach has nothing to do with the law and its function. The recommendation reads: be nice to your neighbor, fellow man, and co-worker and avoid conflict. But the law becomes relevant only in view of conflicts. The point is: who can afford to push his point and how far this can be made independent of local reputation, merit, and exchange gifts and good will. In fact, the only great delegalizer with a minimum of rules and a maximum of effects that has been invented in legal history is the institution of property because of its clear and simple way of predeciding conflicts. Judged against this background, the present problems of overstrain in the legal system are to a large extent consequences of the deterioration of property.

A different theory of the legal system—not only calling it a system but in fact using systems theory as a framework for theoretical developments—will lead to different results. A precise definition of the function of law is essential,[33] otherwise it would be impossible to limit the consideration of functional equivalents or alternatives to the law. Moreover, systems theory requires and offers a conceptualization of structural strain and its sources. Structural strain is a quite normal affair, resulting from the fact that no structure can absorb all problems that emerge in the relation between system and environment. Overstrain simply means the probability of structural change stimulated by too many unsolved problems and overburdened activities.[34]

The theory of open systems that are based on concurring self-reference (or: autopoietic closure) proposes a way of reformulating this problem. All autopoietic systems have to live with an inherent improbability: that of combining closure and openness.[35] Legal systems present a special version of this problem. They have to solve it by combining normative and cognitive, not-learning and learning dispositions. On the screen of the analytic framework of scientific description this requirement may appear as contradiction. In fact, however, a social system can live up to opposed necessities. In the course of its evolution the system hits upon exceptional conditions that permit such combinations if they become incorporated as structural constraints. The system uses incidental chances and makes them work, thus developing by accident. In this way the improbable becomes probable.[36]

From this point of view strain can be conceived of as *residual improbability*, and overstrain as too much of it. Given a certain institutional framework routinizing the improbable there may be a *non plus ultra*. In other words, although we shall feel unable to outline last limits, the institutions show signs of suffering. They show signs of overload and of more or less unsuccessful attempts to solve fundamental problems by insufficient means. "Involution" becomes the predominant reaction to evolution—involution in the sense of progressive complication, variety within uniformity, virtuosity within monotony.[37]

If increasing improbabilities are the problem, neither technological nor communal devices will satisfy. The social engineering approach to the law is a political approach—and of course, completely legitimate as a perspective of the political system. In fact, the increasing differentiation and autonomy of the legal system must entail a relative loss of control over other systems, and the increasingly instrumentalistic view of legal institutions and norms can be understood as an attempt to compensate for this loss of control.[38] The communal approach appears as a countermovement, shifting the power base of the law from central to local pressures and from written instructions to face-to-face interaction. Both converge in negating a proper function of the law as such. Both solve the problem of high improbabilities by transferring it to another system—be it the political system or the countless systems of face-to-face interactions. However, these systems cannot solve it by legal or illegal means; they can only act *extra legem*.

Moreover, the law has to offer protection against reasonable designs and against moral pressures because in an open, post-Gödelian society reason and morality are partisan values.[39] At least, the law has to make sure at which points and how far resistance against demands propagated in terms of a reasonable or moralistic "discourse" might be successful. To maintain this possibility of conflict with reason and morality is one aspect of the differentiation and the improbability of the law.[40]

Given the proper function, the normative closure and the autonomy of the law as constraints, how can the legal system "factorize"[41] its inherent improbabilities? To pursue this question we have to revisit the mechanisms that combine closure and openness. The central device is conditionality. In this respect residual improbability and hence overstrain comes about by using conditional programs for the attainment of ends that are not within the reach of immediate causal operations.[42] Task setting and technology always imply control over some of the causes and lack of control over others.[43] The controlled sector (instruments) may be more or less decisive in relation to the uncontrolled sector. The combination may be more or less arbitrary and contingent. The constellation of causes may be more or less complex. In general, the task is less representative for the unity of the system (i.e., its autopoiesis) if the combination of causes needs a higher degree of contingency and complexity.[44]

There is no apodictic objection against using conditional programs as subroutines in goal programs. The law can very well organize patterns of higher security within result-oriented projects. But including the desired result, in spite of all risks, in the normative framework of the law contributes unavoidably to overstrain, depending on the degree of complexity and contingency of the goal program. To localize and factorize overstrain, a careful task analysis will be helpful. It will reveal many cases in which the law is misused to convey the impression of security where in fact only reasonable guesses can be obtained. Result-oriented legal practice endows opinion with authority. This is a useful device to implement politics by collectively binding decisions. From the point of view of the legal system we have to think of those who lose in litigation and of those who want to invest in legal security. Both will not be served by a legal system that spoils the self-reproduction of normative meanings by conjectural justice.

The way of handling conditionality has an important impact on the sources and patterns of complexity. Until the end of the eighteenth century common opinion attributed the complexity of the legal system to the quarrels of lawyers and to never-ending disputes over the interpretation of law and over problems of legal doctrine. On the other hand, legislation was hailed as the source of simplification, clarification, and transparence of the law. From time to time the formal prohibition of any citing of legal opinions before the court has been considered (and even enacted).[45] Today, the reverse fits the facts: legislation is seen as the main source of complexity, and the quest for order-in-variety is, with less and less hope to be sure, addressed to the general principles of legal doctrine.

This reversal correlates with the increasing differentiation of the legal system and with the increasing stress on learning and non-learning dispositions within the system. It has become irreversible by evolution. However, this is not to say that we have to accept the status quo. Legislation creates complexity because it is at the same time the implementation of policy and result-oriented legal practice. Obviously, result-oriented practice is the most important single source of complexity within the system (in older times, it was litigation and diversity of interests *as such*). Result orientation will, to a large extent, not achieve its ends and will produce unintended side effects. If no-fault divorce increases the rate of divorce or if it changes the bargaining position of husband and wives, was this intended? And if not, what can be done to cope with such results? Such disappointments are fed back into the system and legislation is again its main learning mechanism. Thus, legislation incites legislation. Ecclesia reformata semper est reformanda. Observation of the results of the law means change of the law: the change of conditions conditions the change.

It is difficult to see how legal doctrine can develop in face of such a state of turbulence. Any attempt to compete with legislation on the same level producing a dogmatic set of principles or decision rules will be a futile exercise. Possibly, doctrine merges with legal theory specializing in reflection. Its domain could be the self-observation and self-description of the system. It may produce sober, detached, and "experienced" statements like: no-fault liability means shifting the costs of insurance. This will not immediately slow down the process of change and certainly not contribute to delegalization. It may speed up the process of exhaustion of good intentions, pointing to the fact that the stock of better states is indeed limited. This does not interfere with political responsibility, but it may prevent innovation by comment. But will the legal system and society support a representation of the law that specializes in balanced judgments and lacks commitment to "essential" norms? If this comes out as an unavoidable adaption to overstrain we shall find ourselves no longer motivated to fight for the law or, as Socrates thought to be his duty, to die for the badly applied law.

Endnotes

1. This is again under critical review today. See Richard L. Abel, "Delegalization: A Critical Review of its Ideology, Manifestations, and Social Consequences," *Jahrbuch für Rechtssoziologie und Rechtstheorie* (1980), 6:27.
2. Niklas Luhmann, "Communication About Law in Interaction Systems," in K. Knorr-Cetina and A. V. Cicoural, eds., *Advances in Social Theory and Methodology: Toward an Integration of Micro- and Macro-Sociologies* (Boston, Mass.: Routledge and Kegan Paul, 1981), p. 234.
3. Niklas Luhmann, *The Differentiation of Society* (New York: Columbia University Press, 1982), p. 122.

4. In fact, the most important paradigm change in general systems theory, brought about during the last decade, consists of extending the concept of self-reference from the level of the structure to the level of the elements of the system. This requires a redefinition of the conceptual apparatus of systems theory and shifts its focus from design and control to autonomy, from planning to evolution, from the distinction between statics and dynamics to problems of dynamic stability, etc. See Humberto R. Maturana and Francisco J. Varela, *Autopoiesis and Cognition: The Realization of the Living* (Dordrecht: Reidel, 1980) and Francisco J. Varela, *Principles of Biological Autonomy* (New York: Elsevier, 1979), for a biological version of this approach.

5. Ashby defines cybernetic systems as "open to energy but closed to information and control." See W. Ross Ashby, "Principles of the Self-Organizing System," in W. Buckley, ed., *Modern System Research for the Behavioral Scientist* (Chicago, Ill: Aldine, 1968), p. 4. Another source of this recent interest in combinations of closure and openness is J. Y. Lettvin, Humberto R. Maturana, W. S. McCulloch, and W. H. Pitts, "What the Frog's Eye Tells the Frog's Brain," *Proceedings of the Institute of Radio Engineers* (1959), 47(11):1940.

6. It may be helpful to remember the theological ancestors of *concomitans*, implying the copresence of God in everything that happens on earth. Seen against this background the idea of concurring self-reference reveals itself as a figure of secularization.

7. Niklas Luhmann, *Die Wirtschaft der gesellschaft* (Frankfurt: Suhrkamp, 1988).

8. "Because" in this phrase is no mistake, but intention. It corroborates a point made by Torstein Eckhoff and Nils K. Sundby in "The Notion of Basic Norm(s) in Jurisprudence," *Scandinavian Studies in Law* (1975), p. 123. See also Torstein Eckhoff, "Feedback in Legal Reasoning and Rule Systems," *Scandinavian Studies in Law* (1978), p. 41.

9. See Michel Virally, "Le Phénomène juridique," *Revue du droit public et de la science politique* (1966), 82:5, who specifies a Durkheimian position for *l'ordre judique* as such. Similar opinions use the notion of (social) institution. See Neil McCormick, "Law as Institutional Fact," *The Law Quarterly Review* (1974), 90:102.

10. Further reading will lead to many variants of late transcendental or idealistic positions looking for validity in the realm of spiritual beings. See Heinrich Henkel, *Einführung in die Rechtsphilosophie* (Munich: Beck, 1977), p. 550.

11. Erik Wolf, *Griechisches Rechtsdenken* (Frankfurt am Main: Klostermann; vol. 1, 1950, vol. 2, 1952).

12. "Norm here means, obviously, what people ought to do," writes Paul Bohannan in *Social Anthropology* (New York: Holt, Rinehart and Winston, 1963), p. 284. Also, for Robert Nisbet, "the most vital character of social norms" is the "sense of *oughtness* they inspire in human conduct." See Robert Nisbet, *The Social Bond: An Introduction to the Study of Society* (New York: Knopf, 1970), p. 226. Such formulations may rely on others that state that " 'Ought to be' is a primary, irreducible content of consciousness." See Nicholas S. Timasheff, *An Introduction to the Sociology of Law* (Cambridge, Mass.: Harvard University Press, 1939), p. 68.

13. I follow a suggestion of Johan Galtung, "Expectations and Interaction Processes," *Inquiry* (1959), 2:213, who tries to link normative and factual expec-

tations within a behavioristic framework. For further elaboration see Niklas Luhmann, *A Sociological Theory of Law* (London: Routledge, 1985).

14. Inconsistency, of course, is always a culturally defined fact, and it is well known that advanced societies perceive as inconsistent what primitive societies would treat as regular irregularity. See Peter Winch, *Ethics and Action* (London: Routledge and Kegan Paul, 1972), p. 8. This increasing awareness of inconsistencies can be explained as a result of the increasing differentiation of cognitive and normative orientation.

15. This neologism has been invented, it seems, by Le Moigne. See Jean-Louis Le Moigne, *La Théorie du système générale: Théorie de la modélisation* (Paris: Presses Universitaires de France, 1977), p. 58.

16. See Herbert Blumer, "Ths Psychological Import of the Human Group," in M. Sherif and M. O. Wilson, eds., *Group Relations at the Crossroads* (New York: Harper, 1953), p. 185; Barney Glaser and Anselm Strauss, "Awareness Contexts and Social Interaction," *American Sociological Review* (1964), 29:669; Ronald D. Laing, Herbert Phillipson, A. Russell, and A. R. Lee, *Interpersonal Perception: A Theory and a Method of Research* (London: Tavistock, 1966); Jean Maisonneuve, *Psycho-sociologie des affinités* (Paris: Presses Universitaires de France, 1966, p. 322; Thomas I. Scheff, "Toward a Sociological Theory of Consensus," *American Sociological Review* (1967), 32:32; and Vladimir A. Lefebvre, "Formal Method of Investigating Reflective Processes," *General Systems* (1972), 17:181.

17. See Niklas Luhmann, *Soziale System: Grundriß einer allgemeinen Theorie* (Frankfurt: Suhrkamp, 1984). See also Niklas Luhmann, "Autopoiesis, Handlung und kommunikative Verständung," *Zeitschrift für Soziologie* (1982), 11:366.

18. In a very general sense conditionality is a prerequisite of any complex system that cannot activate all variables at once but has to condition the state of the actualization of others. See Ashby, "Principles of the Self-Organizing System," p. 108.

19. In more "philosophical" terms the reality of law is a process (in Whitehead's sense) consisting of events that for themselves and others combine self-identity and self-diversity. See Alfred N. Whitehead, *Process and Reality: An Essay in Cosmology* (Cambridge: Cambridge University Press, 1929). It may be necessary to say that this holds true not only for legal procedures in the narrow sense but for all events that are communicated (and thereby given unity) with reference to the legal system: for contracts, offenses, births, remarriages, divorces, deaths, etc.

20. Helmut Willke, "Three Types of Legal Structure: The Conditional, the Purposive and the Relational Program," in Gunther Teubner, ed., *Dilemmas of Law in the Welfare State* (Berlin: de Gruyter, 1986), pp. 280–298.

21. Unger also hesitates: "Modern jurisprudence ... increasingly accepted the notion that the meaning of a rule, and hence the scope of a right, must be determined by a decision about how best to achieve the purposes attributed to the rule. But all such purposive judgments are inherently particularistic and unstable: the most effective means to any given end varies from situation to situation, and the purposes themselves are likely to be complete and shifting." See Robert M. Unger, *Law in Modern Society: Toward a Criticism of Social Theory* (New York: Free Press, 1976), pp. 86 and 194.

22. Within the context of the German discussion about the predictive capacity and responsibility of legal decisions my own position is rather extreme—if

not on the right side then on the side of "taking rights seriously." For more balanced views see Gunther Teubner, "Folgenkontrolle und responsive Dogmatik," *Rechtstheorie* (1975), 6:179; Thomas W. Wälde, *Juristische Folgenorientierung* (Frankfurt am Main: Athenäeum, 1979); Hubert Rottleuthner, "Zur Methode einer folgenorientierten Rechtsanwendung," *Archiv für Rechts- und Sozialphilosophie, Wissenschaften und Philosophie als Basis der Jurisprudenz* (1979), special issue 13, page 97; and Gertrude Lübbe-Wolff, *Rechtsfolgen und Realfolgen* (Berlin: Duncker and Humblot, 1981). One hopes that authors who see good chances for the legal control of consequences are not the same as those who increasingly begin to complain about *Verrechtlichung* and overstrain of the law.

23. The distinction between a juridical and a dogmatic level applies to the Parsonian distinction between technical and institutional levels in formal organizations. See Talcott Parsons, "A General Theory of Formal Organization," in Talcott Parsons, *Structure and Process in Modern Societies* (Glencoe, Ill. Free Press, 1960), p. 59. There is also a place for managerial levels in between, materialized, for example, in the form of court policies, organizational policies, intercourt relations, etc.

24. Jan M. Broekmann, "Legal Subjectivity as a Precondition for the Intertwinement of Law and the Welfare State," in Gunther Teubner, ed., *Dilemmas of Law in the Welfare State* (Berlin: de Gruyter, 1986), pp. 76–108; Thomas C. Heller, "Legal Discourse in the Positive State: A Post-structuralist Account," in Teubner, pp. 173–199.

25. See Niklas Luhmann, *Rechtssystem und Rechtsdogmatik* (Stuttgart: Kohlhammer, 1974), p. 49, on the societal adequacy of legal concepts, and see Gunther Teubner, "Substantive and Reflexive Elements in Modern Law," *Law and Society Review* (1983), 17:239, on "responsive law" (I would prefer "responsive dogmatics"); stimulating also Josef Esser, *Vorverständnis und Methodenwahl in der Rechtsfindung: Rationalitätsgarantien der richterlichen Entscheidungspraxis* (Frankfurt am Main: Athenäeum, 1970).

26. Gunther Teubner, "After Legal Instrumentalism? Strategic Models of Post-Regulatory Law," in Teubner, pp. 299–325.

27. Dieter Nörr, *Rechtskritik in der römischen Antike* (Munich: Beck, 1974); Lodovico Muratori, *Dei difetti della giurisprudenza* (1st ed. Venice: 1742; Milan: Rizzoli, 1958).

28. Holy law in particular needs this kind of treatment in view of secondary intentions. See Joseph Schacht, "Die arabische hijal-Literatur," *Der Islam* (1926), 15:211.

29. Based on the legal prescription of the Twelve Tables: "Si pater filium ter venumduit, a patre filius liber esto."

30. Abel, "Delegalization," p. 27.

31. Rüdiger Voigt, *Verrechtlichung: Analysen zu Funktion und Wirkung von Parlamentarisierung, Bürokratisierung und Justizialisierung sozialer, politischer und ökonomischer Prozesse* (Königstein/Taunus: Athenäeum, 1980); Rüdiger Voigt, "Mehr Gerechtigkeit durch mehr Gesetz? Ein Beitrag zur Verrechtlichungsdiskussion," *Aus Politik und Zeitgeschichte* (1981), 21:13; Günther Ellscheid, "Verechtlichung sozialer Beziehungen als Problem der praktischen Philosophie," *Neue Hefte für Philosophie* (1979), 17:37.

32. With slight modifications this small group approach can be extended to the small head group (leading actors) of the "new corporatism." Here, conflict repression by the "holy watching" of well-disposed neighbors and the immobi-

lization by partners can be avoided. But the mechanism of delegalization will be power rather than peace.

33. My own proposal would be: using the possibility of conflict for a generalization of expectations in temporal, social, and substantive aspects—a slight variation of the definition given in Luhmann, *A Sociological Theory of Law*. It is certainly not sufficient to use very general definitions—say, contribution to the order of society—because this would confer on anything the status of being a functional equivalent of the law.

34. Or the probability of "collective behavior"! See Neil J. Smelser, *Theory of Collective Behaviour* (London: Routledge and Kegan Paul, 1963), p. 67.

35. I have to admit that the authors of the theory of autopoietic systems, Maturana and Varela, probably would not subscribe to this formulation. See Maturana and Varela, *Autopoiesis and Cognition*. They underline the necessity of closure and assert that the distinction between system and environment presupposes an observer. Taking this into account we have to postulate that self-observation is an essential characteristic of autopoietic systems. See also Gordon Pask, "Organizational Closure of Potentially Conscious Systems," in M. Zeleny, ed., *Autopoiesis: A Theory of Living Organization* (New York: Elsevier, 1981), p. 265.

36. Edgar Morin, *La méthode* (Paris: Seuil, 1977), 1:294; Niklas Luhmann, "The Improbability of Communication," *International Social Science Journal* (1981), 33:122 (chapter 4 in this volume).

37. See Alexander Goldenweiser, "Loose Ends of a Theory on the Individual, Pattern and Involution in Primitive Society," in Robert H. Lowie, ed., *Essays in Anthropology, Presented to A. L. Kroeber* (Berkeley, Calif.: University of California Press, 1936), p. 99 and Clifford Geertz, *Agricultural Involution: The Process of Ecological Change in Indonesia* (Berkeley, Calif.: University of California Press, 1963), p. 80, who describe involution as increasing the tenacity of basic patterns, internal elaboration and ornateness, technical hairsplitting, and unending virtuosity.

38. See Klaus A. Ziergert, *Zur Effektivität der Rechtssoziologie: Die Rekonstruktion der Gesellschaft durch Recht* (Stuttgart: Enke, 1975), who insists on instrumental and expressive functions of the law. I would prefer to distinguish between political and legal uses of the law.

39. With this (not strictly post-Gödelian but post–French Revolution) evidence in view, authors of the nineteenth century maintained that the function of the law is to create and to warrant freedom. See Georg F. Puchta, *Cursus der Institutionen* (Leipzig: Breitkopf and Härtel, 1856), p. 4, who puts this function explicitly in contrast to the exigencies of reason and morality, i.e., in contrast to Kant. "Freedom," in other words, is the normative counterpart of the fact that a functionally differentiated society cannot base its integration upon the traditional semantics of nature, reason, and morality.

40. By the way, the legal system is not the only one pretending to independence from the supremacy of reason and morality. For politics, read Machiavelli. For love, see the famous "Dialogue de l'Amour et de la Raison," in François Joyeux, *Traité des combats que l'amour a eu contre la raison et la jalousie* (Paris: Hauville, 1667), p. 231. Value-free science is but another version of the same issue.

41. This term is taken in the sense given to it by James G. March and Herbert A. Simon in *Organizations* (New York: Wiley, 1958), p. 191.

42. Helmut Willke, Three Types of Legal Structure.

43. D. J. Mathew, "The Logic of Task Analysis," in Peter Abel, ed., *Organizations as Bargaining and Influence Systems* (New York and London: Heinemann, 1975), p. 103.

44. One of the hermeneutic laws of Schleiermacher reads: "Der Zweck entfernt sich um so weiter von der Idee [=internal unity of the work], je mehr Willkuer in der Produktion ist." See Friedrich E. D. Schleiermacher, *Hermeneutik und Kritik* (Frankfurt am Main: Suhrkamp, 1977), p. 175.

45. For a discussion of this topic see Muratori, *Dei difetti della giurisprudenza*, p. 111.